Routledge Revivals

Agricultural Development in the Mekong Basin

Originally written for the International Bank for Reconstruction and Development, Resources for the Future published this report in 1971 believing it would be of interest to more general readers. Resources for the Future draws on its past experience of resource utilization, development and economics to comment on issues faced by the Mekong Basin River for agricultural development such as the demand-supply conditions for improved agriculture, the limitations imposed by physical and human resource conditions and measures needed to modernize their agricultural methods. This title will be of interest to students of Environmental and Economic studies.

Agricultural Development in the Mekong Basin

Goals, Priorities, and Strategies

Resources for the Future

RFF PRESS
RESOURCES FOR THE FUTURE

First published in 1972
by Resources for the Future, Inc.

This edition first published in 2016 by Routledge
2 Park Square, Milton Park, Abingdon, Oxon, OX14 4RN
and by Routledge
711 Third Avenue, New York, NY 10017

Routledge is an imprint of the Taylor & Francis Group, an informa business

A Library of Congress record exists under LC control number: 70158820

ISBN 13: 978-1-138-94730-6 (hbk)
ISBN 13: 978-1-315-67014-0 (ebk)

Agricultural Development in the Mekong Basin

Agricultural Development in the MEKONG BASIN

Goals, Priorities, and Strategies

A STAFF STUDY

RESOURCES FOR THE FUTURE

Published by Resources for the Future, Inc.

Distributed by The Johns Hopkins Press, Baltimore and London

RESOURCES FOR THE FUTURE, INC.
1755 Massachusetts Avenue, N.W., Washington, D.C. 20036

Resources for the Future is a nonprofit corporation for research and education in the development, conservation, and use of natural resources and the improvement of the quality of the environment. It was established in 1952 with the cooperation of the Ford Foundation. Part of the work of Resources for the Future is carried out by its resident staff; part is supported by grants to universities and other nonprofit organizations. Unless otherwise stated, interpretations and conclusions in RFF publications are those of the authors; the organization takes responsibility for the selection of significant subjects for study, the competence of the researchers, and their freedom of inquiry.

This book results from a staff study undertaken by Resources for the Future for the International Bank for Reconstruction and Development. It was edited and designed by Pauline Batchelder. The map was drawn by Clare Ford.

RFF staff editors: Henry Jarrett, Vera W. Dodds, Nora E. Roots, Tadd Fisher.

Manufactured in the United States of America
Distributed by The Johns Hopkins Press, Baltimore, Maryland 21218
Library of Congress Catalog Card Number 70-158820
ISBN 0-8018-1294-1

Preface

This publication is not one of Resources for the Future's usual type of research monographs, but rather a report made to the International Bank for Reconstruction and Development in response to a specific request. Its primary purpose was accomplished when it was submitted to the Bank. We believe, however, that the report will be of interest to a considerable number of other readers and are therefore publishing it, with only minor modifications. The Bank has advised us that it has no objection to publication. The opinions expressed are, of course, those of the authors.

Members of the group that prepared the report were:

RFF Staff:

Michael F. Brewer
Marion Clawson
Pierre Crosson
John V. Krutilla
Hans H. Landsberg

Others:

Jasper Ingersoll, Catholic University
Nancee Black
Jeffrey Vaughan

Marion Clawson and Hans Landsberg took the lead for RFF in organizing the work and bringing the parts together in the final report.

Joseph L. Fisher

JANUARY 1971

Contents

Introduction

Destruction and disorder in the part of the world that formerly made up Indochina has not stood in the way of rising interest and activity aimed at harnessing to productive, peaceful purposes the flow of the Mekong River, which traverses this theater of warfare. This circumstance gives room for hope that when the war in Vietnam and the associated conflicts in the neighboring countries have come to an end, a long period of constructive development will ensue, making full and intelligent use of the river and its tributaries. Until hostilities cease, many of the sites which offer potential for development will be too insecure for construction or even for surveying; local governments will be too preoccupied with problems of the present to devote major attention to the more remote future; and indigenous populations will be subject to displacement and harassment. Populations and governments will be too critically concerned with survival to permit much thought—let alone action—for long-run development. Thus, planning, studies, experimentation, data collection, and the like will continue to take up a substantial portion of available funds. So far about \$55 million out of total "operational resources" estimated at just short of \$200 million available or pledged between 1957 and the end of 1969 have been so employed.

Perhaps 1957 best marks the beginning of recent planning efforts for the Mekong Basin. It was in that year that the Bureau of Flood Control and Water Resource Development of the United Nations Economic Committee for Asia and the Far East (ECAFE) produced a report entitled "Development of Water Resources in the Lower Mekong Basin." Not only did it greatly enlarge and deepen the analysis made in a brief study produced by ECAFE in 1952, but the discussion that it engendered in the Thirteenth Session of ECAFE led to establishment of the Mekong Committee,[1] a regional group that has ever since been the driving force in all enterprises associated with Mekong River development. Organized at the suggestion of ECAFE, it is made up of Thailand, Cambodia, Laos, and South Vietnam, the four riparian countries of the stretch of the river that constitutes the Lower Mekong (which begins roughly where the river turns east after having hugged the Burmese border). They have held continuous membership in the committee, regardless of the state of political relations between them. In the course of 12 years, 26 other countries have contributed to the scheme, supplying funds, equipment, and personnel. Nearly half the funds have come from the riparians, about 15 percent from the United States.

In 1962 the Mekong Committee authorized the preparation of a more ambitious plan than that of 1957. The resulting document, a preliminary draft, known as the Amplified Basin Plan Report,[2] was completed in 1970. It contains the grand design for the next 30 years. Because it projects a construction program the cost of which is put at nearly $8 billion for river-associated facilities alone, omitting complementary investments, because of its magnitude, and because some of the proposed facilities have long lead times and thus call for decisions at an early stage, the Amplified Basin Plan Report opens a major new stage in the planning for Mekong River development.

1. More properly called the Committee for Coordination of Investigations of the Lower Mekong Basin and alternatively referred to as the Mekong Coordinating Committee.

2. Reference is to the *Report on Amplified Basin Plan: A Proposed Framework for the Development of Water and Related Resources of the Lower Mekong Basin,* Draft, June 1970 (Committee for Coordination of Investigations of the Lower Mekong Basin), hereinafter referred to also as ABP Report. It should be emphasized that this document is a draft to serve as a focus of discussion. It is thus indicative of possibilities but in no way a blueprint ready to be implemented. At the same time it is the most elaborate and comprehensive document of its kind to emerge since the Mekong Committee was first established.

One may surmise that it was in recognition of this situation that in the summer of 1969 U.N. Secretary-General U Thant proposed to the International Bank for Reconstruction and Development (World Bank) that it play a greatly increased role in the area and make available its experience and judgment, and presumably also its ability to raise and supply funds. The World Bank responded affirmatively. A special staff unit was formed whose first—and rather formidable—task was to familiarize itself with the available material and the "state of the arts" regarding development proposals for the Basin. It has subsequently made its first concrete suggestions for new initiatives and may be expected to play a key role in the future.

Early in 1970 the Bank approached Resources for the Future with the idea that formation of the Bank's ideas might benefit from an independent study undertaken by an outside group such as RFF that would look at development strategy and priorities in that area, especially as they might affect agriculture. With the exception of a senior staff member who had been a member of a survey team organized by the Ford Foundation in 1961, RFF had no particular experience in the region. Although a handicap in one sense, this lack seemed to argue in its favor in another. It was thought that it would minimize the weight of any acquired prejudices and allow a fresh look at a highly complex situation: not familiar with the bulk of the complexities, such a group might be able to focus on the forest rather than on the trees. It was thought wise, nonetheless, to strengthen the team, and an anthropologist with extensive experience in the Mekong River Basin was added to RFF's staff for the duration of the study.

What follows should be read in that light. It involved no travel to the area, but is based on the vast literature that has by now accumulated on the subject. While it benefited from the knowledge of people familiar with the area—especially in government and international agencies—this benefit again was second-hand. The study did, however, draw on the knowledge and experience of RFF staff members in matters of resource utilization and development in other parts of the world and on their general background in economics of land, water, and energy resources.

On the whole, however, analysis was directed to broad strategy rather than to detailed programming, to ideas rather than to quantification. Moreover, thoroughness of coverage was sacrificed to speedy delivery, since the Bank's staff specified the late fall of 1970 as a necessary deadline. Thus the study is best characterized as a broad look at "where we go from here."

To summarize briefly, the report begins by assessing the demand-supply conditions for an expanded agriculture in the Mekong River Basin, proceeds to evaluate what limitations are imposed by physical and human resource conditions, specifies the measures that are needed to achieve conditions favorable to modernizing agriculture, and inquires into what criteria might serve to establish a sequence of development.

Within this rather conventional-looking scheme, unconventional attention is given to the human and institutional setting in which development takes place. The reason for this is that "progress" is equated not only with increased output or a growing stock of physical facilities but also with a rising ability of the people in the region to gain a measure of material security—via a rising income—in a social context that would secure for them widening choices in matters of production, consumption, location, and life styles generally. The questions are not therefore merely what the river could help to produce and how that production could be utilized, but also—and prominently—what the aspirations of the region's people are, what pace and character of development would be least destructive of their system of values and beliefs, and how the latter could be gradually altered to become consonant with the requirements of modern agriculture.

Asking these questions makes it harder to assign specified roles to funding than if one clung to discussing development of physical facilities. In the latter case there is no lack of precedents. Dams or power plants have been financed before, and the services flowing from them can be well defined. But there are few if any precedents for promoting the kind of changes in people's attitudes and practices that are given prominence in this study, changes without which one is likely to build neither operating facilities that yield satisfactory economic returns nor societies that provide increasing measures of welfare to their constituents, but only "monuments to man the prodigal."[3]

The emphasis placed on nonphysical elements, usually subsumed under the heading "institutional infrastructure," thus derives from the conviction that provision of physical facilities alone cannot be relied upon to trigger those changes that human beings must make in individual behavior and social arrangements. Its implications are twofold. To the investor, the promoter, the developer, it means that he must be alert to the human and social setting in which physical facilities operate—this is the environment to which his plans must be adapted. Awareness of this fact of life will keep him from overestimating engineering and economic efficiency pure and simple. But it also raises the very difficult question of what can be done as a matter of deliberate choice, as opposed to adaptation, to achieve the necessary transformation of values, beliefs, attitudes, and institutions.

Throughout the course of the study the RFF group became more and more impressed by the scope of the enterprise. The area affected by the course of the river and its tributaries measures nearly a quarter of a million square miles. It comprises practically all of Laos and Cambodia, two-fifths of South Vietnam, and one-third of Thailand. Close to 30 million people now live in the area of the watershed involved. That figure is likely to double by the end of the century. The potential for power generation and irrigation is large: the proposed Pa Mong hydro station alone would eventually have a capacity of four to five million kw., more than double the current capacity of Grand Coulee and six to seven times that of Kariba. Water controlled by it could irrigate at least two, and perhaps five million acres. And that is only one project, albeit one of the two biggest. Along the river's 2,600 miles that lie within the Lower Basin the Mekong Committee has identified locations for no fewer than 17 possible projects on the main stem and in the Delta and some 80-odd projects on the tributaries. As of mid-1970, three projects on tributaries had been completed,

3. Asian Development Bank, Tokyo, *Asian Agricultural Survey* (Tokyo, 1969).

TABLE 1. Portions of the Four Countries in the Basin

	Cambodia	Laos	Thailand	S. Vietnam
Population:				
National	6,701,000	2,893,000	34,783,000	17,866,000
Basin	6,239,000	2,701,000	12,803,000	6,892,000
% of national population in the Basin	93	93	37	38
Area (km.²):				
National	181,035	236,800	514,000	173,263
Basin	163,797	220,500	189,029	72,029
% of national area in the Basin	91	93	37	42

Source: Mekong Committee, *Annual Statistical Bulletin*, 1969.

at least with regard to power generation, for a total installed capacity so far of 33,000 kw. and an eventual capability of supplying water to harvest crops from about 150,000 acres. Less than 3% of the total cultivated area in the Basin is now reached by irrigation projects of any kind, and total power generation is a miniscule fraction of the potential.

Thus it is clear that development (1) will affect a large area and a large population, and (2) has barely begun. The major decisions lie ahead; there is time to reconcile differences and to plot new paths. This is not a bad time to be consulted, and RFF has appreciated the opportunity given to it by the World Bank to participate in the process.

1

Development Objectives and Some Major Themes

DEVELOPMENT OBJECTIVES IN THE MEKONG RIVER BASIN

Most of the some 30 million people in the Mekong River Basin[1] are engaged in agriculture, which over the years has adapted itself to conditions in the Basin—the monsoon, the varied topography, the fluctuations of the river system, and, in mutual interaction, the settlement patterns. Agriculture promises to remain a major activity within the Basin, and although it will continue to be constrained by geographical features and the weather, it can be substantially modified to the benefit of the local population. Seed and other inputs better adapted to the environment can increase yields; roads, dams, and canals can overcome some of the limitations of location and water regimens; appropriate organizational changes can extend and stabilize markets for agricultural commodities.

Intervention in the Basin designed to promote development must be guided by objectives toward which the effort is directed and in terms of which its success can be gauged. A diversity of economic, political, and social objectives suggest themselves, not all of which are mutually compatible. This report assumes that the principal objective of development efforts is to increase the material well-being of the people living in the Basin; put somewhat differently, to increase the real per capita product of the region and achieve a more equal distribution of income. These joint

1. Here, as elsewhere in this study, "Mekong" is understood to refer to the Lower Mekong only.

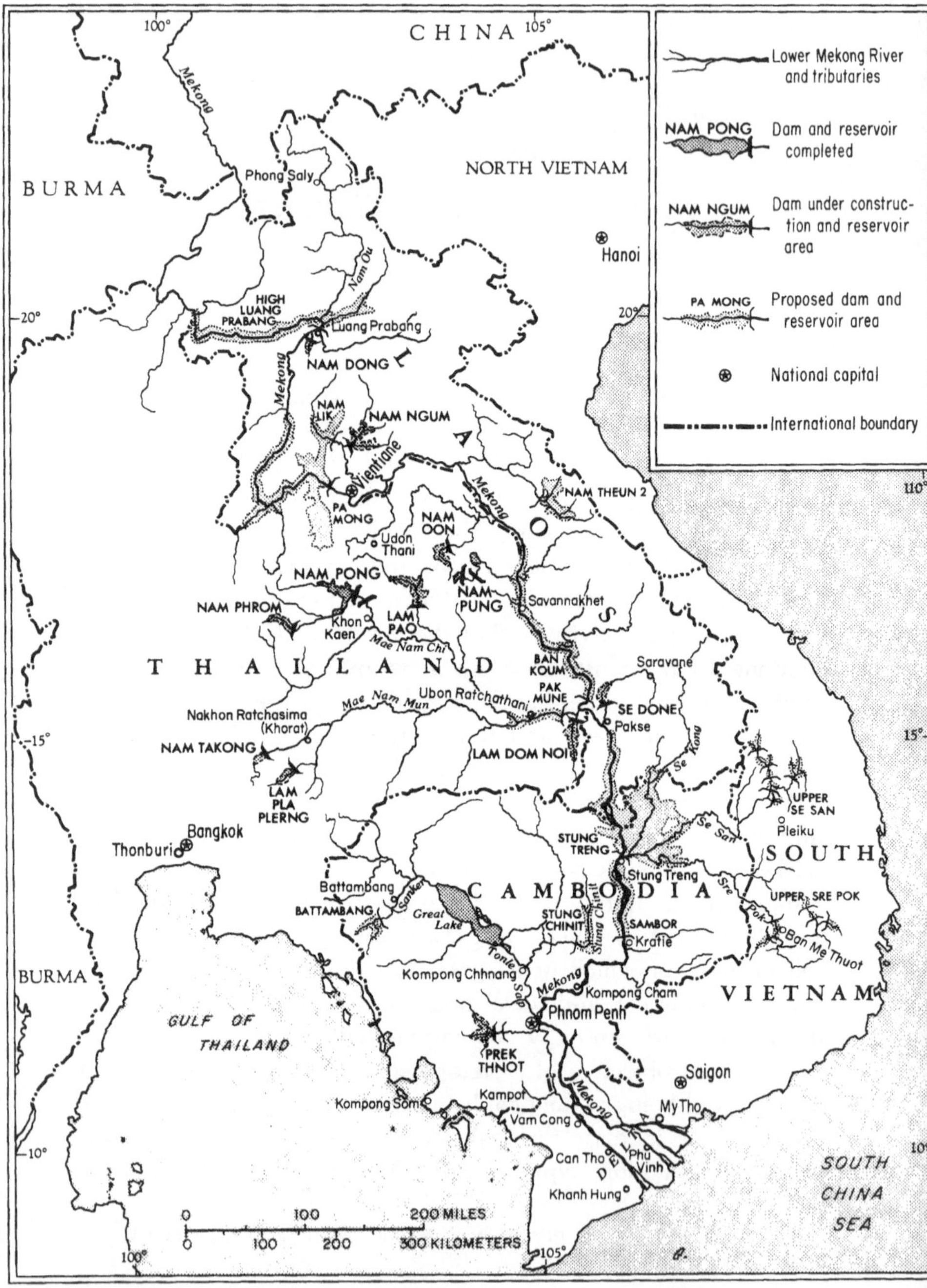
CHINA
BURMA
NORTH VIETNAM
Hanoi
Phong Saly
Mekong
Nam Ou
HIGH LUANG PRABANG
Luang Prabang
NAM DONG
LAOS
NAM LIK
NAM NGUM
Vientiane
PA MONG
NAM THEUN 2
NAM OON
Udon Thani
NAM PONG
NAM PUNG
NAM PHROM
Khon Kaen
LAM PAO
Mae Nam Chi
Savannakhet
THAILAND
BAN KOUM
Saravane
PAK MUNE
SE DONE
Pakse
Ubon Ratchathani
Mae Nam Mun
Nakhon Ratchasima (Khorat)
NAM TAKONG
LAM DOM NOI
LAM PLA PLERNG
Se Kong
UPPER SE SAN
Pleiku
Se San
Bangkok
Thonburi
STUNG TRENG
Stung Treng
SOUTH VIETNAM
Battambang
BATTAMBANG
Sanker
Great Lake
CAMBODIA
STUNG CHINIT
Stung Chinit
SAMBOR
Kratie
Sre Pok
UPPER SRE POK
Ban Me Thuot
Tonle Sap
Kompong Chhnang
Mekong
Kompong Cham
BURMA
Phnom Penh
GULF OF THAILAND
PREK THNOT
Saigon
Kampot
Kompong Som
My Tho
Vam Cong
Can Tho
Phu Vinh
DELTA
Khanh Hung
SOUTH CHINA SEA
0 100 200 MILES
0 100 200 300 KILOMETERS
100°
105°
110°
20°
15°
10°
Lower Mekong River and tributaries
NAM PONG
Dam and reservoir completed
NAM NGUM
Dam under construction and reservoir area
PA MONG
Proposed dam and reservoir area
National capital
International boundary

objectives of efficiency and equity at times may be inconsistent—some per capita real product may have to be sacrificed if certain levels of distributive equity are to be attained. We do not purport to prescribe the most desirable mix of efficiency and equity, but assert that there will have to be sufficient distribution of gains in real economic product so that social cohesion and public confidence in the development effort are sustained.

Several derivative objectives can be identified. For example, gains in material well-being are sought to provide individuals broader opportunities for self-development in whatever directions they may choose. This implies that there should be a strong "social component" to economic development. Not only must skills of individuals be enhanced and new social institutions devised if economic development is to be a self-sustaining process, but development activities must continually foster the capacity of individuals and groups within the region to determine the directions in which they wish to develop and their ability to govern themselves toward that end.

Gains in real per capita product may be sought through increases in regional product, a reduction of population growth, or a combination of the two. The primary emphasis in discussions to date of Mekong River Basin development—and an emphasis retained in this report—is on increasing economic product. However, an important criterion for development activities is that efforts to expand real product of the area not inhibit existing social and economic incentives to reduce the rate of population growth. Wherever possible, activities to expand the Basin's economic product should provide incentives favoring such a reduction. Arrangements for project operation, maintenance, and local financing may offer opportunity to introduce such incentives. For example, if a project provides water supplies, electricity, or some other service the consumption of which is generally proportional to household size, repayment schemes which charge larger households more than smaller would provide a tangible "reward" for smaller families. Similarly, the need for a large number of offspring to support a couple in their old age would be reduced to the extent that their financial security is increased. If new farms are being established as part of a development activity, their size might be established with this in mind—that is, they might be large enough to provide a farmer and his wife sufficient income to set some aside for their old age.

An additional criterion for development activities has to do with long-run threats of environmental disruption associated with these activities. Within the Mekong River Basin, development is likely to entail substantial modifications in the natural regimen of the river and its tributaries, which, given the size of several proposed projects, would have extensive ecological consequences. They may be consequences of the direct impact of projects—for example, the spread of schistosomiasis that might be induced by the creation of canals and reservoirs; or of indirect impacts—the movement of farmers displaced by reservoirs to highlands with soils vulnerable to erosion, and the still insufficiently known effects of irrigation on Basin soils. Although it is unlikely that these consequences can ever be predicted with full accuracy, sufficient resources should be allocated to keep the probability of unforeseen ecological damage low, and to build into projects specific measures to mitigate such damage.

These objectives and criteria are generally in harmony with each other; to the extent that differences arise they will have to be resolved by groups within the Basin and parties to the development effort. It is assumed that assistance from countries and institutions external to the Basin is required to enable the countries concerned and the Basin's residents to achieve these objectives; but the major effort must be mounted by the people of the Basin, and they must consider any programs undertaken as theirs.

SIGNIFICANT MEKONG DEVELOPMENT INVOLVES SUBSTANTIAL CHANGE

Increasing human welfare by introducing new modes of production, new factors of production, and new organizations for supporting these tangible innovations is an evolutionary process. Groups external to the society undergoing development can play a strategic role. They are essential sources of information, technical advice, and capital, which are necessary but are far from the sufficient conditions needed to initiate and facilitate the processes of economic development. These external engines of change must be harnessed by the people who ultimately will direct their own development. This will require departures from traditional behavior at the level of the individual, and his society can acquire the ideas, skills, and competence needed to further the development process only by introducing new social institutions and organization.

Whatever goals may be sought, any development program involves changes, in the course of which some people acquire greater capacities for the production and use of wealth. Since the level of such capacity is embedded in a people's entire social organization and cultural outlook, real development entails considerable social and cultural change. As indicated in Chapter 5, Basin farmers can raise their productive capacities only by profoundly altering their conventional agricultural practices. The Amplified Basin Plan Report stresses change in farmers' production. In our view, it places too little emphasis on the needed changes in marketing to raise farmers' actual income through sale of their crops, changes that must reach beyond the village agricultural sector to the complex set of economic and social institutions and cultural values which currently link village life with the rest of the riparian countries' economy and society.

It is precisely because development involves such extensive change in people's lives that the process of planned development in the Mekong requires full commitment of governments and active participation of people in the Basin, to be achieved by the reorienting and strengthening of governmental and commercial institutions through which government officials, merchants, and village farmers are related, as well as by setting up new *networks* of voluntary organizations.

"Commitment" and "participation" seem big and vague terms. Their meaning here is quite specific: they entail substantial modifications in the way in which the major participants perform their roles, such modification, in turn, flowing from basic alterations in the current system of material and cultural rewards, so that officials can profit from being more honest and efficient, merchants from being more scrupulous, and farmers from responding to the exacting demands of modern agriculture. With this understanding, transformation as well as strengthening of institutions (for governing, banking, extension, irrigation, education, health, marketing, and the like) must be viewed as essential to improving the delivery of inputs and marketing of outputs of modern agriculture sufficiently to produce real and sustained advance in the standard of living of the masses of Basin peoples.

Planned development of a large region commonly involves the introduction of specialized organizations and, equally commonly, their failure to coordinate with each other and with older, indigenous institutions in the region. Such failure has already limited the performance of some Mekong projects associated with tributary development. Given the immense variety of special functions to be filled, and given the limited coordinating responsibility thus far accorded the Mekong Committee, this administrative failure could prove disastrous to major Mekong projects. The issue basically is the need for coherent, coordinated action among local village institutions, regional and national governmental agencies, and agencies of the United Nations and donor countries. The fact that the Mekong Committee has so few historical precedents from which to learn is ample indication that the issue of lack of coordination should be faced openly as a major potential obstacle to the achievement of real human well-being through Mekong development. To facilitate the achievement of well-being, periodic assessments of development activities undertaken to date should be scheduled, with particular evaluation made of the degree to which activities by groups external to the Basin are synchronized with those of the national governments of the riparian countries and other indigenous groups.

SEQUENCE OF TRIBUTARY AND MAINSTEM PROJECTS DESERVES CONTINUING APPRAISAL

Two interrelated issues arise regarding tributary—or subbasin—and mainstem development: relative priorities of small and large projects, and the extent to which the Basin as a whole can be viewed as a single unit of development. The Amplified Basin Plan Report states two criteria used in formulation of the plan:

1. Each component should contribute to the satisfying of definite needs.
2. The implementation of any project should not preclude development of other projects that may be required at some future time.

We would add two more criteria:

3. The construction and operation of each project should maximize local and national participation in order to expand and improve the current technical and managerial skills of the riparian countries concerned.
4. Projects should be selected, designed, and managed with as much consideration given to equitable diffusion of benefits as to efficiency of benefits in relation to cost.

Application of these four criteria and evidence before us argue for favoring intensive development of as many tributary—or subbasin—projects as possible in all four countries as well as flood and salt water control works in the Delta before construction of mainstem projects is undertaken.

As suppliers of power, the mainstem projects appear to become increasingly attractive the longer their construction is delayed past the early 1980's. On the basis of realistic assumptions regarding fuel prices and rates of return to investment, thermal power generation now and for perhaps two decades appears to have the edge over hydro installations on the mainstem, quite apart from any considerations that might arise from taking a closer look at the diffusion of benefits that would flow from mainstem hydro generation.

There is every reason to believe that the modernization of existing agricultural practices in rainfed as well as irrigated areas in conjunction with simultaneous development of subbasin projects would meet food needs of the riparians for the next two decades with enough left over to meet export demand. If inadequacy of skills and management, broadly understood, prevents this goal from being reached in projects associated with tributary development and elsewhere on operating farms, there is no reason to believe agricultural production based on mainstem development would accomplish it. As the outlook for early high returns from mainstem irrigation projects is judged less certain and the need for benefits becomes less urgent, so does early development of the Pa Mong and Stung Treng projects (see map).

Unless there are demonstrable scale economies associated with larger irrigation projects, mere size in terms of hectares irrigated establishes no preference; and, given the scarcity of administrative and water management skills in the regions that these projects would serve, it is in any event doubtful that any benefits of scale economies would soon be realized. By contrast, the prior development of projects in subareas is likely to promote cumulative progress that can set the stage for later mainstem development. Furthermore, the costs of the mainstem projects are so large—certainly in terms of the riparian countries, and probably also in terms of available external funds—that most other lines of investment would be precluded. Moreover, after two decades of studies and expenditure of nearly two hundred million dollars, relatively early payoff is needed to sustain momentum and motivation. Such payoff can be expected only from smaller projects.

It appears unrealistic to expect the irrigation benefits of Pa Mong or Stung Treng to be forthcoming soon after construction is completed. To build the system is one thing, to learn to manage it properly, another. The latter could require 25 years or more. Smaller, subbasin development efforts would make an earlier and probably greater social and educational contribution per dollar invested. In view of the capital required for Pa Mong, its construction should not be initiated until the subbasin projects which precede it have met with demonstrable success.

Criteria of flexibility and hedging development investments suggest that it is preferable to spread the risk of failure over a larger number of smaller projects even if their total expected payoff—in conventional terms—is no greater, or perhaps even less, than what it would be from a smaller number of larger investments. The detrimental impact of one or more huge failures on the immediate users, the farmers, on the organizations supporting them, and on the attitudes and expectations of the whole of rural society could be considerably more severe than the cumulative effect of a number of smaller failures which at least would have contributed to the enhancement of the skills of the farmers, of the providers of inputs and services, and of the recipients of outputs that serve them directly and indirectly.

The relationship between tributary and mainstem development raises the related issue of the extent to which the Mekong Basin is a *unit*. In terms of political alignments, economic interdependence, basic social organization, language groups, cultural values, and ecology the Basin is highly diversified. Nor do the Basin peoples, although most of them have attained basically similar rice cultures and adhere to either of the two major forms of Buddhism, see themselves as occupants of a single area. The issue here then is not one of a common *region* but one of a common *unit for development.*

The Basin does comprise a single hydrological system of interrelated flows of water from the head of the mainstream and every tributary down to the Delta. It is the systematic nature of these water flows, in interaction with each other and with the land forms along the river banks, that presents a unique setting for development by the four national occupants of this otherwise diverse valley. If upon this natural hydrological system were superimposed a physical system of river basin management, each component dam, as an element in the system, would have incremental effects in concert with other dams. As elements of a system, a set of dams would combine to produce a cumulatively greater effect on the possible levels of flood control, low water flow, salt intrusion control, irrigation and drainage, hydroelectricity, and river transportation. The effects of the whole system of dams would be markedly greater than the sum of isolated, individual parts. This possibility of cumulatively greater results from a system—which would apply to a set of tributary dams, mainstem dams, or both together—is the major impelling reason for the four riparian countries to consider the otherwise diverse region of the Mekong Basin as a single unit, or single system, for development.

It is beyond the scope of this analysis, but within that of amplified basin planning, to compare the potential gains and costs of such a basin system (or of subbasin systems) with other forms of development each country might attempt to carry out individually. Any comparison, for example, of a Mekong project with an alternative investment a riparian country might make should include both the predictable results of the Mekong project itself and any benefits resulting from its interaction with other components of a Mekong system. A fundamental assumption of the Mekong Committee is that the four riparian countries can achieve more development by collaborating in a basin system. While this likely is true, the social, institutional, and ecological difficulties of constructing such a system are so formidable that economic results will probably be more fully achieved by prior concentration on subarea development. Planning studies for mainstem projects can simultaneously be undertaken, but construction should be postponed until general development in the less demanding tributary setting is of sufficient substance and scope to promise satisfactory economic and social returns from projects of Basin-wide scope.

FLEXIBILITY IS ESSENTIAL IN ANY PROGRAM FOR CAPITAL INVESTMENTS FOR MEKONG BASIN DEVELOPMENT

Intervening in the Mekong River Basin, a large area of diverse physical and social characteristics, involves substantial uncertainties. Initiating economic processes new to the area entails changing the behavior of individuals and groups whose susceptibility and response to change are not known. The sensitive political situation within the Mekong River Basin suggests that some of the specific objectives of development and the constraints to which development effort is subject may themselves change in unpredictable directions and with unknown speed. Mekong development efforts must, therefore, be so conceived and executed as to afford sufficient flexibility for exploiting opportunities for purposeful change and avoidance of impasses.

Paucity and unreliability of information are among the reasons a flexible development program is needed. It may be possible to reduce the need for flexibility by increasing the amount of information available and its reliability through a systematic program of research initiated early in the development effort. The degree of availability of funds and time will limit the opportunity to do this, but the establishment of an evaluation procedure to assign relative reliability of the information gathered about the Mekong Valley should be made an initial development activity. (See next section.) This is especially true with respect to social, ecological, and economic conditions.

One method of achieving flexibility is to design and launch early development efforts involving projects that embrace relatively small, self-contained areas. In this way, development benefits will accrue when physical circumstances, social attitudes, and motivations within only a small and reasonably homogeneous area are favorable. Often these features are present in physically smaller and less expensive projects, so that more projects may be undertaken within a given budget; and this in turn provides an opportunity for assessing relative performance and learning from such an assessment more about the obstacles and opportunities for development within the Basin as a whole—in short, learning by doing, or by small failures, if necessary.

Another important way in which flexibility can be introduced into a development program is by undertaking investments designed to change the social infrastructure so that the mobility of productive factors is increased, traditional relationships are loosened, and the opportunity for specialized production is extended. Transportation improvements, a standardization of productive factors, and perhaps more reliable alternative sources of factor supply and market information will create more options for economic activity and diversity within the Basin. Economic objectives in a development effort are furthered through opportunities for specialization in production and trade with other areas. Development of product markets and standard grades of products produced will greatly facilitate the capacity of localities to exploit any comparative advantages they may have in production specialization. In addition to these direct economic advantages, broadened markets for both factors of production and final products increase the exposure of individuals and communities to information about their comparative economic situation within the Basin as a whole. Such widening opportunities for "shopping around" favor the acceptance of new ideas, new economic roles, and new productive technologies.

Flexibility will also result from undertaking at early stages projects which are multipurpose in character and capable of being managed in ways which can change the relative amounts of the types of benefits they yield. Mixed farming, for example, is an activity that calls for attention under this heading.

In any of these undertakings an adequate pool of managerial talent within the service area is of utmost importance, to the point of being itself a most important source of flexibility. Managerial capacity is important if the opportunities for more efficient economic activities are to be realized. This suggests that some investments undertaken early in the development effects should be of an educational nature—focusing on primary lines of production and also on the administration of local development efforts. Chapter 4 points to areas needing attention.

GUIDING DEVELOPMENT OF MEKONG BASIN REQUIRES CONTINUOUS COLLECTION AND INTERPRETATION OF DATA

The need for a broad information base is especially critical in the Mekong River Basin. Termination of military hostilities in the area and political settlement may well have an economic development component which calls for some large, visible investment with a Basin-wide impact. But whatever event or succession of steps creates such a situation, if the development program within the Basin is to be essentially agricultural in character—and given the nature of the Basin it is hard to see how it can be otherwise—the information yielded from a careful evaluation of a spectrum of different types of subbasin projects is a very important if not crucial factor in success for the larger mainstem projects which will follow.[2]

Most development efforts in the Mekong River Basin will have to contend with limited or inadequate information. It undoubtedly will be necessary to supplement existing data through field investigations, yielding results that are likely to be uneven in both coverage and reliability. Such investigations are expensive in time, talent, and dollars and should contribute to a deepened general understanding of development problems and processes as well as being useful for coping with particular problems immediately at hand. The reliability of field investigation results must be assessed if these results are to be used by a substantial number of people confronting a diverse set of development problems. The likelihood of achieving considerable economies of scale in evaluating field data and research suggests that the task be undertaken by some type of regional information center. Various development agencies could be served if such a center periodically published substantive abstracts of available studies in a general format. This would greatly facilitate the evolution of a reasonably complete physical profile of the Mekong River Basin.

2. A focus on electricity would lessen the opportunity for securing information on Basin-wide development, but even here several tributary projects would entail hydroelectric generation and could yield information on matters such as the following: How is power assimilated into the community? To what extent and where is it used to expand economic production? To what extent does it constitute a consumption item to support conditions of relative luxury for a small percentage of the population who enjoy high incomes and live in urban areas? To what extent and through what paths does it serve to raise the level of living more generally and equitably?

Social data, however, often cannot be secured in such a straightforward fashion. There are two general categories of social information that are important for development planning: 1) descriptions of existing social institutions and characteristic behavior patterns; and 2) information on behavioral changes resulting from some purposeful development effort or activity. Social data of the first type can be secured through surveys and investigations by sociologists, anthropologists, geographers, and others. Many such studies have been undertaken in the Mekong Basin. Their utility would be enhanced by the evaluation and abstracting procedure suggested above. The second category of social information must be secured by systematically designed studies to observe reactive behavior. The results of such studies contribute to a deeper understanding of the social attributes of the indigenous society, but they also provide an indication of how well certain types of development activities perform within that social context. Indeed, each such investigation can be considered as a miniature "controlled experiment." The early phase of a Basin development program should give relatively high priority to designing appropriate experiments of this type, and the early investment schedule should include an array of carefully monitored projects that are representative of the principal development alternatives available. The collection of this kind of social information would be greatly facilitated by a periodic review procedure, in which development progress within the Basin is systematically evaluated. This assessment might benefit from being performed under the auspices of a major multinational group, such as the World Bank or the Asian Development Bank.

2

The Importance of Agriculture

Agriculture is the dominant economic activity in the Mekong Basin and no doubt will remain so for many years to come. Data on production and employment in the Basin alone are scarce. In the four riparian countries as a whole agriculture accounts for from 30 percent of Gross Domestic Product in Thailand and Vietnam to considerably more than that in Cambodia and Laos. Since the major urban centers, Bangkok and Saigon, lie outside the Basin, one must infer that the share of agriculture in production is substantially higher in the Basin than in the countries as a whole. Output per man in agriculture is only a fraction of what it is in nonagricultural employment, and the labor force in agriculture is higher than agriculture's share of total production—ranging from at least 65 percent in Vietnam to 80 in Laos and Cambodia.

With agriculture ranking so high in output and employment, it follows that agricultural development is the key to general economic development in the Basin. This is not to deny that the growth of towns and cities within the Basin and the development of an infrastructure that binds urban and rural parts together is extremely important; but growth in total economic output and improvement in the welfare of ordinary people will be nearly impossible without major change in agriculture. This report, accordingly, is devoted primarily to consideration of the region's potential for agricultural development and to the conditions necessary for achieving that potential. Other lines of development, such as generation and use of hydropower, are dealt with mainly as events that impinge upon the sequence of investments.

The potential for agricultural development in the region depends upon the growth of demand for commodities which are, or could be, grown there and upon the ability of the region to compete in these expanding markets. Rice is by far the most important single commodity product of the region at present. It accounts for about one-half (Thailand) to two-thirds (South Vietnam) of total agricultural production, occupies 55 to 80 percent of the arable land, and no doubt employs comparable shares of the farm labor force. The prospective growth in demand for rice inevitably is of major significance for the region's agricultural economy.

Rice prices have been under considerable downward pressure for the last couple of years. In the judgment of both international agencies, such as the Food and Agriculture Organization, and national ones, such as the U.S. Department of Agriculture, the long term prospect apparently is for more of the same. World demand can be expected to increase slowly, probably not more than 2.5 to 3.0 percent annually, while the introduction and rapid spread of the improved varieties offers the possibility of sharp increases in production. Hence, the outlook is for strong competition among the various producing countries. The competitive situation is clouded by the fact that a number of traditionally importing countries (India, Indonesia, Pakistan, the Philippines, and Malaysia, for example) have adopted policies of self-sufficiency in rice and apparently are prepared to pay a high price to achieve this. A further complication is the uncertainty concerning the future role of Burma, the world's leading rice exporter as recently as 1963 but now a poor fourth behind the United States, Thailand, and Mainland China, and of Mainland China itself, by far the world's leading rice producer. No one knows the potential for expansion in China, but the absolute volume of the country's productive capacity is so enormous that even a relatively small percentage increase could have a major impact on world markets. (In 1967–69 Mainland China accounted for almost one-third of world rice production. In 1968 it was the second most important exporter.)

Under these circumstances it probably would be unwise for the countries in the region to count on foreign markets to absorb much of the increase in rice production that can be anticipated over the next couple of decades. This is not as serious a constraint as might be thought, since exports presently take only a small proportion of total production. (Even in Thailand only 14 percent of total production was exported in 1965–67.) Moreover, the domestic demand for rice in these countries could easily expand at 3.0 to 3.5 percent annually for the next two decades. (Say 2.5 to 3.0 percent on account of population growth and 0.5 percent reflecting per capita income growth of 2.5 percent and income elasticity of demand of 0.2. The prospective decline in prices also ought to provide some small stimulus to demand.)

If this picture of the demand prospects for the region's rice production is correct, then nonrice production will have to grow at about 5.0 percent annually for the region to achieve satisfactory growth in total income.[1] The prospective expansion in demand for the region's nonrice commodities appears sufficient to support this rate of increase in production. If regional income grows at the target rate of 5–6 percent annually, then local demand for fruits, vegetables, meat, and dairy and poultry products could easily increase at about the same rate. The income elasticity of demand for these commodities in the region is quite high, in some cases exceeding 1.0, according to FAO figures. If

1. This statement is based on the following considerations:

a. Regional population will grow at 2.5–3.0 percent annually over the next decade or so.

b. Per capita income growth targets will be 2.5–3.0 percent. Hence total income must grow at 5.0–6.0 percent.

c. Agricultural production accounts for about 40 percent of total regional output. If the total income growth target is set at 5.5 percent and nonagricultural production grows at 6.5 percent (not unreasonable), then total agricultural production must grow at about 4.0 percent annually.

d. Rice accounts for about 60 percent of total agricultural production. If rice production grows at 3.25 percent annually, then nonrice production must grow at about 5.0 percent to achieve the growth target for total agricultural production of 4.0 percent.

the region manages to expand livestock production in response to rising market opportunities this will also greatly stimulate local demand for feed grains.

External demand for the region's nonrice production also appears favorable over the next decade. Thailand already has established itself as one of the world's leading exporters of feed grains, mostly corn to Japan, and there is no reason why it should not maintain its position in this expanding market. High and rising incomes in the developed countries promise considerable growth also in world demand for tropical fruits and vegetables.

Thus it is not unreasonable to expect demand for the region's agricultural production to grow at roughly 4.0 percent annually over the next decade or so. If production increases in response to this growth in demand, the performance of the agricultural sector will be consistent with the achievement of overall income growth targets of 5–6 percent. However, the necessary rate of increase in agricultural production is not likely to occur automatically. In fact, the rate is likely to fall well short of that needed unless productivity in agriculture increases markedly. As noted, foreign demand for tropical fruits and vegetables probably will increase considerably over the next decade or so; however, competition from producers in Africa, Latin America, and other parts of Asia in serving these markets will be intense. To maintain its position or increase its penetration, Mekong region agriculture must meet standards of price, product quality, and reliability of supply which are yet beyond its capacity.

Competition from foreign producers in serving local markets will of course be less severe, particularly since some measure of protection for threatened products can be expected. Without substantial improvements in productivity, however, the price paid for protection could be extraordinarily high. Moreover, increased productivity is the only source of rising per capita income and welfare in agriculture, surely a major objective of development policy in the region.

The productivity increases needed are not confined simply to operations at the farm level. Rather, the entire structure of agriculture, viewed as a system of activities embracing not only the farmers themselves but also the institutional and physical infrastructure providing supply and marketing services to them, must be markedly improved. The conditions necessary for achieving these improvements, therefore, are of major importance in development planning for the region. They are taken up in the succeeding two chapters.

3

Utilization of Physical Resources in the Basin

The previous chapter noted the key role of agriculture in development of the Mekong region, and it emphasized that successful fulfillment of this role would require substantial increases in agricultural productivity. Rising productivity results from the substitution of modern for primitive production techniques. The process of technological substitution involves a complex set of changes resulting in (1) employment of greater quantities and better qualities of the physical resources needed for agricultural production, and (2) improved capacities of people to use those resources. This chapter attempts to discern the bottlenecks presently inhibiting the employment of both more and better quality resources, while the subsequent chapter considers the social and cultural framework within which capacities in resource management could improve.

Despite the scarcity and unreliability of rural labor force and employment data, there is a consensus that at present there is some measure of both unemployment and underemployment among farm workers. Given this surplus labor and the prospects for rural population growth, the region is not likely to suffer from a shortage of bodies available for employment in agriculture. There is still rather considerable potential for expansion also in the land base. But simple quantitative increases in employment and cultivated area will not do the job. The quality of the labor force and the productivity of the land must also be improved considerably. Hence the region must make substantial investments in public health and educational services available to farmers, and it must take measures also to assure greatly increased and reliable supplies of economically priced fertilizers, pesticides, improved seed varieties, animal stocks, feed, farm machinery, and other inputs essential to modern agriculture. Moreover, water supply systems must be improved so that farmers can exert much greater control over the timing and amount of water applied to their fields.

At present the productivity of both land and labor resources in the region is low because of the generally primitive technology employed. As shown in Table 2, fertilizer consumption is at a small fraction of the levels achieved in Asian countries which have made rapid progress in agriculture.

According to the ABP Report[1] the level of use of pesticides and farm machinery in the Basin also is very low in relation to levels found in countries which have successfully launched programs of agricultural modernization. Similarly, the improved rice varieties so far have not been widely adopted in the region. A recent FAO report[2] states that in the late 1960's the new varieties were of no importance in Thailand, Laos, and Cambodia, while in South Vietnam they were sown on about 200,000 hectares. (This would be slightly more than 10 percent of the rice area in the Delta region of Vietnam. However, what percentage of these 200,000 hectares was in the Delta is not known.)

The reasons for the failure of farmers in the Basin to adopt a more modern technology are complex, but the main features of an explanation can be discerned. It is well to begin, perhaps, by dismissing one argument that has been advanced from time to time, although with decreasing frequency in recent years. This is that farmers remain locked in a primitive technology because of the force of tradition and prejudice and from lack of interest in producing more, even when they clearly could do so by adopting more modern practices. Substantial worldwide evidence indicates strongly that this argument is fallacious. The evidence shows that when the risk-discounted payoff to more modern techniques is clearly above the yield of primitive technologies, peasant farmers

1. Annex V-9, pp. 1-2.
2. *Recent Trends and Patterns in Rice Trade and Possible Lines of Action.*

TABLE 2. Fertilizers Consumed in 1965–1966

(kilos per ha. of cultivated land)

	N	P_2O_5	K_2O
Vietnam	9.0	18.9	3.8
Thailand	1.6	0.9 [a]	0.4 [a]
Taiwan	162.9	41.9	51.0
South Korea	89.1	42.2	17.8

Source: Asian Development Bank, Tokyo, *Asian Agricultural Survey* (Tokyo, 1969).
Note: Data for Laos and Cambodia not available.
[a] 1964–1965.

will rapidly adopt the more modern practices. The performance of Mekong Basin farmers shows a similar responsiveness to economic opportunity.[3] Lao farmers have responded to vegetable market opportunities; Thai farmers to kenaf and corn opportunities; Cambodian farmers to new opportunities with cash crops; and Vietnamese to various opportunities for fresh foods and fibers. Moreover, Basin farmers have demonstrated a uniformly positive reaction to the prospect of irrigation water. Comments by on-the-spot observers bear this out. So does an economic benchmark survey, part of the Pa Mong feasibility study, which reported that the great majority of Lao and Thai farmers were quite prepared to pay fees for irrigation water despite the fact that neither government normally charges such fees in irrigation projects. The farmers anticipated that irrigated farming would be difficult, and over 90 percent of both the Thai and Lao farmers responding to the survey said they would be willing to go to a district training center in order to receive special training for such cultivation.

Hence the persistence of primitive production techniques in the region cannot be ascribed to farmers' lack of interest in economic opportunity. The more likely explanation is that more

3. See, for example, J. R. Behrman's study of Thai agriculture, *Supply Response in Underdeveloped Agriculture* (Amsterdam, 1968), esp. p. 337; also R. Muscat, *Development Strategy in Thailand: A Study of Economic Growth* (New York, 1966).

modern technologies have not been available on economically attractive terms. The reasons for this can be traced to weaknesses in the physical and institutional infrastructure serving farmers in the region. The fact is that this infrastructure is not currently capable of supplying farmers with the input and marketing services required to implant and sustain a modern technology. Perhaps the most glaring weakness in this respect is in the water management systems commonly employed throughout the Basin, and in this context we refer to water from *any* source, not just that originating in irrigation projects. Thus, water management, as pointed out below, is of importance not only to the relatively moderate portion of land considered "irrigable," but easily to more than half the land. To tap the full production potential of improved seed varieties, fertilizers, pesticides, and other "modern" inputs requires a degree of water control well beyond what most farmers now can achieve. Not only is water generally in short supply in the dry season; but present systems fail to provide farmers with the necessary degree of control when water is available. Only 2.0 to 2.5 percent of the cultivated area is "under the command of irrigation projects." Of the rest, at least one-half and probably much more is dependent upon natural inundation.[4] Under this system the farmer has little control over either the total amount of water applied to his fields or the distribution of it to various parts. Water not retained flows onto neighboring fields, and so on—needlessly draining fertilizer in the process—until the surplus, if any, is taken off by means of a canal or drain.

Weaknesses in the water management systems employed in the Basin go far to explain the failure of farmers to adopt the new rice varieties on a larger scale. These varieties require precise control of water depths if maximum yields are to be obtained. At the time of transplanting, the water should be moderately deep—

4. ABP Report, p. II–46. The statement that one-half of the cultivated area in the Basin is dependent upon natural inundation seems to conflict with other data showing that much more than 50 percent of the cultivated land is in paddy. The normal practice is to water paddy land by natural inundation. Hence it seems likely that well over half of the cultivated area is watered in this fashion.

perhaps 5 cm. Ideal water depths gradually decline for the next 45 to 50 days, until the tillering stage is past; then they should rise to 5 to 10 cm. during the stages of panicle formation, heading, and flowering; after which the water should be fully drained away during the ripening stage. This degree of precision is very difficult, if not impossible, with existing water management systems.

Apart from demanding more precise control of water depth and timing of application, the new rice varieties also require closer control of water *quality* than the traditional varieties. This is particularly important in those parts of the Delta subject to salt water intrusion. As salinity rises, crop yields fall. Until adequate fresh water flows can be maintained in rivers and canals in the Delta, the use of high yielding rice varieties, especially during the dry season, will not be economical in much of that area.

Though close to 500,000 acres of riceland in South Vietnam were planted to the new varieties in 1969 and more than twice as much was planned for the current year,[5] the role of existing water management systems in inhibiting the spread of the new rice varieties is widely recognized by those concerned with the region's agriculture. In a paper given at the Regional Seminar on Agriculture sponsored by the Asian Development Bank,[6] a Thai official attributed the lack of interest among Thai farmers in the new varieties to "many limiting factors including the cooking quality, the short stature, the fact that they are not suitable for rainy season culture and the high cost of nitrogen fertilizer." These are, of course, factors that are being successfully dealt with elsewhere and that have, in fact, led to the breeding of more attractive strains, apart from the fact that greatly higher yields are an important argument in their favor. More significantly, he went on to say that year-round multiple cropping is only in the pilot project stage and that "little will be achieved until further

5. See U.S. Department of Agriculture, *Foreign Agriculture,* Oct. 12, 1970.

6. *Regional Seminar on Agriculture, Papers and Proceedings* (Seminar held in conjunction with Second Annual Meeting of Board of Governors of Asian Development Bank, Sydney, Australia, April 10-12, 1969), p. 212.

improvements in irrigation and drainage systems can be brought to the farm level."

At the same conference, spokesmen for Cambodia and Laos also emphasized the inadequacy of present irrigation systems as the principal obstacle to the introduction of the improved varieties of rice. (Of 2,500,000 hectares sown to rice in Cambodia in 1967 only 100,000 were irrigated. Projects "under consideration" would add another 138,000.)

Ambitious plans for raising new varieties on one-third of South Vietnam's ricelands in 1971 are questioned in a recent U.S. Department of Agriculture article: ". . . it is questionable whether land conditions and water resources in areas not already planted to improved varieties are exploitable."[7]

The Development and Resources Corporation, in projecting the expansion of rice production in the Vietnam delta over the next 20 years,[8] considers three sources of increased production: (1) extension of the cultivated area; (2) improved practices applicable to traditional rice varieties; (3) movement in several stages toward double-cropping of the new, high-yielding varieties. These are expected to contribute 55 percent of total rice production by 1989. Achievement of this goal, however, will require major programs to improve water management: irrigation, flood control, drainage, and protection against salinity intrusion.

Poor water control systems also help to explain why Basin farmers have not made greater use of fertilizers. This is attested to by the Asian Development Bank's *Asian Agricultural Survey* (1969), which asserts that "lack of water control, resulting in either excess or inadequate water can sharply reduce the yield response to fertilization. This accounts for the fact that rice farmers in the region rarely use fertilizer on rice grown under rain-fed or flood irrigation conditions, even when fertilizer is available."

7. *Foreign Agriculture,* Oct. 12, 1970.

8. Development and Resources Corporation, *Projected Agricultural Production: 20-Year Mekong Delta Development Program* (New York, 1969).

The weakness of presently employed water management systems is an important factor inhibiting the spread of modern agricultural techniques in the region, and strengthening these systems is indispensable for achieving sustained technological progress in agriculture in those areas in which either supply or disposal of water, from local precipitation or from remote sources, represents a problem. Improvement in water management systems will not be sufficient, however, for there are other important bottlenecks in the physical and institutional infrastructure serving farmers. The rural transport system is inadequate to provide farmers quick and economical access to input and product markets. Too many roads, for example, are poorly constructed and maintained, so that the cost to farmers of reaching these markets is greatly increased. Storage, processing, and machinery repair facilities and skills are in short supply. The net effect of these various bottlenecks is that farmers cannot be sure that they will be able to purchase various "modern" inputs where, when, and in the amounts needed, nor that, should they purchase these inputs, they will be able to market the larger volume of production yielded by them. In the absence of such assurances, incentives to adopt more modern practices perform poorly.

For a variety of reasons, irrigation tends to receive the lion's share of attention in attempts to improve the performance of agriculture in less developed countries, and certainly in the Mekong River Basin countries. Water is a basic necessity for growing crops; the unutilized river flow presents a highly visible challenge; the methods for harnessing it are well known; and the new grain varieties require large amounts of water. For the opposite reasons, improvements in rainfed agriculture usually receive low priority. However, no single factor stands out as a sufficient key to change. Rather it is a combination of new practices to be observed rather rigorously and a break with many beliefs, firmly implanted in the cultivator's mind, that opens the way to higher yields.

Water management plays an important part in unirrigated farming as well. As for rice, dry season rice production is at present unknown in most of the Mekong Basin. In the Delta, perhaps 30,000 acres of rice have been grown in the wet season in the past. Lack of water during the dry season has made rice production impossible at that time. However, it is possible that rice yields in the dry season would be higher than in the wet season, given the same varieties and the same other inputs. During the wet season, there is a great deal of cloud cover, and the amount of sunlight falling on the land is reduced; during the dry season, sunlight is much greater, and with the careful use of fertilizer and other factors, high yielding rice varieties might utilize this additional sunlight to produce higher yields.

The main importance of unirrigated farming, however, lies not with rice but with other farm enterprises. There has been some clearing of forest land for unirrigated farming in the past decade, especially in northeast Thailand. Part of the increased corn production in Thailand has been on such land, though northeast Thailand accounts for no more than about 10 percent of the country's corn crop. Kenaf, on the other hand, is practically all grown in this area, and the acreage has in the last 15 years risen to about 300,000 hectares. These are nearly all uplands, not subject to water overflow and not suitable for rice production; a more limited increase in acreage of these two crops has occurred on the higher or better drained land within areas that are generally given to rice growing.

The climate presents substantial difficulties in rainfed agriculture. Rainfall is erratic in onset and volume, tends to be intense, and is subject to heavy runoff. Moreover, it comes in two segments, interrupted by a dry spell. Water supply control thus offers large benefits. Its absence severely restricts choice of crops, achievement of high yields, and sustained levels of output.

The transition from shifting to settled agriculture is a prerequisite for the introduction of modern farming practices and the achievement of rising incomes. The shifting type of cultivation on much upland, mentioned above, raises several questions about the future. How much more land of this kind might be cleared and farmed? How much is this dependent upon a forest-clearing–burning–farming–abandonment rotation? What kinds of fertility and erosion problems have resulted so far or might result? How far may it be possible, through adaptation of research done elsewhere, to overcome the fertility problem through use of fertilizers, or by different cropping sequences, or in other ways, and what would be the cost in relation to achievable yields and anticipated prices? What conservation practices would maximize the utilization of water, and what role would be played by fallowing and pasturing? In terms of markets, corn production has the great advantage that a ready export market exists, primarily in Japan. But how certain can one be that domestic markets for other cash crops might develop rapidly, even provided continuity of both supply and demand were established?

Answers to the questions raised above are not readily apparent. To the extent that additional lands can be brought into production and that lands recently cleared and in production can be kept continuously in crops, the opportunities for increased agricultural output in Thailand are considerable. Apparently less has been done along this line or can be done in the other riparian countries. In all of them, the need for experimentation on a pilot project scale is urgent.

The spread of more modern practices is inhibited also by weaknesses in the institutional infrastructure serving farmers. Surveys made in the region indicate that the absence of credit or its high cost has seriously limited the greater use of fertilizers and other capital inputs. Most farmers lack the resources to purchase these inputs in the amounts needed to take full advantage of the improved rice varieties. Hence, even if water management systems permitted economical employment of these inputs, most farmers would be unable to purchase them. Short term credit on reasonable terms is essential. Yet the rural areas throughout the region are very poorly served by banks or other credit institutions, the principal credit sources being so-called "noninstitutional lenders"—small merchants, traders, relatives, or friends. Frequently the rates charged are very high, and in any case the total supply of funds is low in relation to the need if purchases of fertilizers and other inputs are to approach the amounts required to exploit fully the potential of the new varieties.

The institutions providing input and marketing services to farmers also are weak. This may in large measure simply reflect the weaknesses in the physical infrastructure, however. If the transportation network were improved and storage and processing facilities expanded, it is possible that private initiative would quickly provide the input and marketing services needed to support a more modern agriculture. The speedy reaction of Thai middlemen in responding to the opportunities created by the Friendship Highway supports this view.

Suppliers of modern agricultural inputs are generally moderate-sized Chinese family firms located in provincial and district towns. Recently, however, international firms with some capital ownership and control by nationals have begun to enter this field. A few of these firms are prepared to invest in staff development efforts in order to reach farmers with modern agricultural information as well as supplies and equipment. Most local Chinese firms, however, cannot take such a long-term development perspective.

There are, finally, some companies owned by senior government officials, such as the various government monopolies in Thailand, that combine Chinese capital and managerial skills with the political power and some capital provided by the national elite.

Perhaps the most important commercial institution providing marketing services to Basin farmers is the rice mill located in or near almost every town. Basin farmers, particularly in Thailand and South Vietnam, now sell increasing amounts of rice and other cash crops. The rice millers, who, like input suppliers, are predominantly Chinese, have come to perform a multitude of functions for the farmers. Rice millers extend credit during the rice season, buy the crop at harvest, transport the crop to their mills, provide some short-term regional storage of rice, and sell it locally or to the national and international markets. Because they are usually Chinese and are economically much stronger than the farmers, the rice millers are generally feared and suspected. The millers nevertheless have long-established economic relationships with farmers in their area. The functions they currently perform, and additional ones they might perform in the modernization of agriculture, are worthy of intensive on-the-spot investigation and perhaps of becoming the substance of pilot projects.[9] Although Chinese merchants in Southeast Asia generally have very extensive commercial links with each other, relatively little is known about the actual socioeconomic relations which town rice millers have with national millers. Despite the sensitivity of ethnic relations, a network involving rice millers might become a most effective and inexpensive way to supply farmers with some of the necessary inputs for modern agriculture and with the means to market their rice crops.

Among the modest beginnings of industrialization in the Mekong Basin, probably the most important thus far are the agricultural processing plants. They provide part time employment to farmers and boost the national economy through the much higher export prices fetched by processed or partly processed products.

The land tenure systems prevailing in the Basin also affect the abilities and incentives of farmers to shift to more modern production techniques. Farms throughout the region are generally very small, too small in many cases to justify purchase of a tractor. This need not seriously inhibit mechanization *provided* machinery services can be made available to small farmers on a custom basis. While this practice is not unknown in the region, it has not yet attained a scale sufficient to reach most farmers.

The small size of farms is not nearly so serious an obstacle to greater consumption of fertilizers. In fertilizer use, as opposed to mechanization, there are few if any economies of scale. The same is true also of improved seeds and pesticides. Hence, if the various other obstacles to more widespread use of these inputs could be overcome, and custom machinery services were made available, farm size would not seriously impede the adoption of modern production techniques.

This is not to say, however, that the prevailing land tenure systems do not inhibit technological progress in Basin agriculture.

9. See also p. 86.

Land tenure involves many aspects of the ownership, control, and utilization of the land apart from size of holdings, and these affect the abilities and incentives of farmers to invest in new practices.

No uniform or comparable figures for landownership exist for the Basin in all four countries. Absentee landownership is a serious issue in the rich Delta area. In the Vietnamese delta about 70 percent of the farm families are primarily dependent on cultivating rented land for which they pay the landlord the equivalent of about one-third of their crops. In a series of measures begun in 1969 the Saigon government has initiated a program to give land to the cultivators with fair compensation to the owners. This program, if carried through, would have far reaching effects on South Vietnamese rural life. In the plateau and valley areas farther upstream, absentee landownership is not yet a serious issue.

This contrast between the Delta and the Basin upstream (like the contrast between northeastern Thailand and the more fertile central valley of Thailand) highlights a serious difficulty: as irrigation and other development projects make land a more profitable investment, outside investors are attracted to it. The high rents and uncertainties about absentee landlords, however, deprive tenants of adequate security or incentive to invest in expensive modern agricultural inputs in order to obtain the potentially higher yields. Substantial absentee landownership has also contributed to widespread agrarian discontent, as so intensely displayed in Vietnam, to which it has in the past been confined. But an ominous portent is the current trent toward accumulation by townsmen, in both northeastern Thailand and Laos, of tracts of land near irrigation projects under construction, in the expectation that they will soon become much more valuable. This trend has not yet brought about a serious level of landlessness in the Basin above the Delta, where most farmers own some land and many farmers rent additional land as they have manpower to work more than their own holdings.

A second major difficulty regarding landownership is that of differing interpretations and degrees of formal ownership. A recent sample study of Lao and Thai farmers in the Pa Mong region indicated that none of the Lao farmers held any written title or claim to their land; they would have to pay an unofficial fee for the survey and the document. Among the Thai families none held a full title deed; a certificate of land use covered 22 percent of the land owned, a statement of occupation covered 57 percent of the land owned, a tax receipt covered 5 percent, and claim to the remaining 16 percent rested on an informal claim of land use without any written documents.[10] Informal land use claims or less formal papers may suffice for relationships with village neighbors and with local administrative officials, but they prove wholly inadequate for determining land boundaries and compensating owners at times when confiscation may be necessary for development projects. Lacking such formal documents of ownership, farmers become embroiled in disputes and can suffer major losses from land confiscation.[11] Different interpretations of ownership can often lead to loss of land and security by farmers. The effects of such uncertainties in weakening farmers' incentives to innovate are obvious.

Government institutions relating to rural areas may have an important impact on the pace of technological change in agriculture. Such institutions are of two basic types: those promoting development and those enforcing law and order.

Government development institutions are relatively new, although they have proliferated greatly in less than two decades. While tending to maintain social distance between themselves and villagers, development officials have in considerable numbers gone to villages and met with farmers. Their numerous programs of assistance to villagers have helped somewhat not only to narrow the urban-rural gap, but also to reduce village institutional strength

10. U.S. Department of the Interior, Bureau of Reclamation, *Pa Mong Stage One Feasibility Report* (1970), p. VII–6.

11. Lao farmers have reported a few cases of rich families in the Nam Ngum area (where the largest of the current crop of projects is now under construction) recently informing cultivators of ancient titles to land which the cultivators had presumed to be their own.

by attempting to perform some of the functions once handled by village institutions.

Government institutions devoted to the maintenance of law and order have far older roots in rural areas than have development institutions. The district office, presided over by a district chief or district officer, is generally the lowest unit of paid, full-time civil service in the Basin countries. The major agent of the national government to the people in the locality, its dominant traditional duties have been those of maintaining order.[12] Generally the most unpopular unit of government has been the police forces, whose actions in rural villages throughout the Basin have earned suspicion, fear, and unspoken resentment. Too often these hostile feelings have been transferred to development programs, thus weakening some relatively constructive but unprecedented and uncertain efforts.

12. Some development programs are of genuine interest to village farmers, but since many recent development activities have been carried on by units dispatched from the more remote provincial government or central government, they tend not to be related to each other or to the local district office.

4

Characteristics of Human Resources in the Basin and Their Implications for Development

The capacity of a people to make the best use of available physical resources involves their technological knowledge and skills, but more broadly, the very fundamentals of their understanding of life and of their social and cultural organization. The focus in this chapter is on demographic and settlement patterns, manpower resources, cultural values, and social institutions as these factors relate to people's abilities to use the physical resources of the Mekong Basin for agricultural development. A secondary emphasis here is on the systems of material and nonmaterial rewards among the Basin peoples and the links between these rewards and agricultural development.

The variety of ethnic groups and the levels of population density within the Mekong Basin are very high and are closely interrelated. The highest concentrations of people occur in the various subbasins most suitable for paddy rice cultivation, while fewer than four people per square kilometer inhabit the uplands of northern Cambodia and eastern Laos. The relationship between ethnic groups and population density is a constraint on Basin development. First, the national ethnic majority peoples (Lao, Thai, Cambodian, Vietnamese) occupy the most favorable agricultural lowlands in all four countries. Second, the Chinese residents are clustered as merchants in the towns and as truck gardeners and merchants in villages near the towns. Third, Vietnamese penetrated long ago as farmers, fishermen, and merchants into Cambodia; more recently as merchants into Laos; and as truck gardeners and fishermen into parts of Thailand. Fourth, the highlands on the edges of the Basin are thinly populated by ethnic minority groups such as the Cham and upland Khmer in Vietnam; the Cham, upland Khmer and upland Thai groups in Cambodia; and upland Thai along with small Sinitic minorities such as Meo and Yao in Thailand and Laos. Living in uneasy relationships in different ecological niches and at different economic levels, these varied peoples present real difficulties for general Basin development.

The populations of the Mekong Basin are extremely young: about 60 percent of the population in each of the four Mekong countries is under 25 years of age. In both absolute numbers and percentage terms young children and old people tend to be clustered in the villages while adolescents and young adults have moved into the cities and towns—an important feature in development terms.

Population growth has been prodigious: between 1920 and 1970 Cambodia grew from over 2 to nearly 7 million, Laos from less than 1 million to 3 million, Thailand from about 10 to about 35 million, and South Vietnam from perhaps 5 to about 18 million. On the basis of estimated population growth between 1963 and 1968, the Mekong Committee computed the following annual growth rates: Cambodia 2.2 percent, Laos 2.4 percent, Thailand 3.1 percent, and Vietnam 2.6 percent.[1]

Cities of over 20,000 people have been growing much faster than population generally—from 5 to 10 percent annually in the riparian countries. The capital cities especially have increased at enormous rates, though the growth of Vientiane and Saigon has been accelerated as a result of insecurity in the countryside. The 1970 population (before the war spread to Cambodia) in the four capital cities, in rounded figures, is estimated as follows: Vien-

1. A frequently made but frequently forgotten caution regarding the poor quality of statistics for the Mekong region should be repeated here for demographic and manpower statistics. Many of the people in the Basin who gather and report statistics have neither the concerns nor facilities for precise compilation of data. Readers with both such concerns and such facilities must keep this difference in mind. The statistics reported here, based upon the Mekong Committee *Statistical Bulletin*, are probably adequate to indicate general orders of magnitudes and general comparisons among the four riparian countries.

tiane 150,000, Phnom Penh 500,000, Saigon-Cholon about 2 million, and Bangkok-Thonburi nearly 3 million.

Each of the four capital cities is several times larger than the second largest city in each country. Apart from these capital cities, of the remaining forty-nine urban centers in the Basin only two have over 100,000 population, and only eight have over 50,000. All of these are in Vietnam or Thailand.

The farming villages of the Mekong Basin vary widely in size but tend to be mainly of two types of settlement pattern. In Laos, northeastern Thailand, and Cambodia outside of the floodplain they are mostly compact clusters of households on slightly higher wooded land in the midst of rice fields, averaging about 50 households in Laos and varying from 100 to about 1,000 households in northeastern Thailand and Cambodia. In the Cambodian and Vietnamese delta the most frequent settlement pattern is that of a linear village strung along the sides of canals and rivers. The villages are usually composed of more than a thousand households and in Vietnam usually consist of several component hamlets, each one with its own hamlet chief. Although composed of several hamlets, usually the Vietnamese village is the more self-sufficient unit, with its village council, school, one or more temples, shops, and the like.

Levels of population density in the Mekong Basin do not yet seem very high by riparian standards and certainly not in comparison with levels elsewhere in Southeast Asia. Despite the limitations of comparative statistics and comparative impressionistic observation, both approaches suggest that general levels of living in the Basin are comfortable by Asian standards.

The rates of annual population growth, however, are high by general Asian and world standards, and this causes one to question whether the people living in the Mekong Basin can double their population from nearly 30 million to over 60 million in the next three decades and still produce enough additional goods to raise their standard of living substantially. A second and more particular concern derives from the fact that the concentration of population is greatest in the subbasins most suitable for wet paddy cultivation. Thus, being the most populous, the most irrigable areas will have to increase their productivity by the greatest amount if they are to continue supporting their inhabitants.

Even more serious are the interactions of rapid population growth and Mekong development. The 1970 ECAFE review of the social situation in Asia points out that many Asian governments have given increased emphasis to population control as their countries have fallen short of development targets set earlier. But government programs, ECAFE comments, thus far have been too narrowly focused upon the techniques of fertility control without adequate regard for the general demographic, economic, and social conditions which have caused accelerated population growth. Providing adequate economic security for urban people may still be impossibly difficult for underdeveloped economies. Some increase in economic security for rural people may, however, be within reach. This could substantially reduce pressure for growth in rural Basin populations and thus ease the problems of the agricultural development process itself.

Rapid urban growth in the four capitals in the Basin may also seriously limit Mekong development. Outside the Vietnamese delta most of the provincial and district towns have relatively small populations, less than 30,000. Clear evidence on this matter is as yet lacking, but it may be questioned whether towns of this size can provide the many urban functions necessary in the introduction of broad scale agricultural Basin development. The concentration of urban institutions and facilities within a very few urban centers probably tends to limit development-associated modernization throughout the countryside.

An important element in this limitation is the tendency for youth and young adults to cluster in the towns and cities, leaving the villages populated largely by children and older adults. This diminishes rural receptivity to modern practices and perhaps creates greater economic dependence of villagers on relatives in towns.

Large village societies usually possess greater capacity to develop the social institutions required to channel development activities. The varied sizes of agricultural villages are likely to affect Mekong agricultural development. In particular, the small villages in Laos will be at a disadvantage in creating institutional structures of needed types and scope.

MANPOWER CHARACTERISTICS

The most important single way of measuring or characterizing human resources in the Mekong Basin is in manpower terms. An adequate statement of manpower conditions of the Basin would include data about the following items and relationships among them: current and projected numbers of people with their types and levels of occupational training; current and projected types and levels of training needed to implement Mekong programs; current and projected educational and training programs; current and projected levels of investment in these educational and training programs. Relationships among these items would provide the basis for planning future manpower programs in the Basin. To be sure, the above scheme of data is probably not fully obtainable for any part of the world. In the Mekong Basin, however, such data are very fragmentary for Thailand and Vietnam and almost totally lacking for Laos and Cambodia. The task of manpower assessment facing the Mekong Committee is formidable.

The priorities and composition of the Mekong Committee have placed further constraints on manpower studies thus far. A reading of the Amplified Basic Plan Report indicates clearly that in gathering data and in planning future levels of investment for manpower resources and physical resources, the Mekong Committee has thus far given far more emphasis to physical resources of the river basin than to the current and future resources of its people.

Complete or even partial implementation of the short and long term aspects of the Amplified Basin Plan—or whichever modified version will eventually emerge from the current draft—will require improved quantity, quality, and location of skills over what has been previously envisaged. It may be true that each member country of the Mekong Committee can meet initial requirements for trained manpower "to greater or lesser degree" as the ABP

Report claims.[2] Yet the quantity of middle and upper level manpower demanded, as well as the geographical and chronological dimensions of that demand, are not analyzed in specific relationship to the timing, size, and regional distribution of investment in physical works. To facilitate policy making a general manpower survey based on a common analytical method and employing consistent definitional categories is needed.[3] If manpower planning is left to the separate national governments and concerned international organizations without any attempts being made to achieve uniformity, the different methods and the conflicting results deriving from their application which now characterize the available manpower data will be perpetuated. For future "Basin planning" to be effective, a mutually satisfactory method of analyzing manpower requirements should be agreed upon, and the assumptions regarding the timing and pattern of Basin investments should be integrated into the separate national plans if total *national* manpower requirements are to be compatible with *Basin* requirements. This is especially true for Thailand and Vietnam, the two countries with regions located outside the Basin demanding trained manpower. A comprehensive Basin manpower and education survey and planning project, assisted jointly by the International Labor Organization and the United Nations Educational, Scientific, and Cultural Organization and carried on by the Mekong Committee for all four riparian governments would perhaps be the best approach.

THE RURAL EMPLOYMENT PROBLEM

The theoretical and empirical difficulties involved in the concepts of disguised unemployment or underemployment in agriculture are well known.[4] Their discussion here has narrow limits,

2. p. V–185.

3. To date, the only study of this sort has been UNESCO's *Higher Education and Development in Southeast Asia* (1967). The work of the ILO Asian Manpower Team in 1971–73 should be helpful in this regard.

4. Gunnar Myrdal, *Asian Drama: An Inquiry into the Poverty of Nations* (New York, 1968), Vol. II, pp. 959-1027.

since no attempt has been made, to our knowledge, to measure the degree of rural "underemployment" in Vietnam, Cambodia, or Laos. For Thailand, information exists on urban unemployment and average number of working days per farm family by month,[5] but it is doubtful that these data would justify a discussion of the need for shifting the "surplus" to other occupations.[6]

However, development targets set solely in terms of rates of growth in per capita income often fail to bring about acceptable growth in productive employment or do so only with a considerable lag. Often the conflict between maximizing income and maximizing employment is reflected in the conflict between the desire for productive capital-intensive investment in the industrial sector and the need for employment-creating investment in the agricultural sector.

The problems of underemployment or unemployment argue for development planning giving employment creation a high priority. In the long run a successful policy of population control may serve to ameliorate the problems implicit in capital scarcity and high population growth rates. In the short run, however, productive use must be made of available labor to accelerate economic growth. This implies that labor should be substituted for scarce capital when economically feasible.

The broad-based population pyramids of the riparians suggest that unemployment can become a serious problem if youth cannot find employment in the agricultural sector and upon migration to urban areas discover fewer job opportunities than expected. On

5. Thailand, Government National Economic Development Board, *Fact Book on Manpower in Thailand* (Bangkok, 1967). Ian Bowen, *Seven Lectures on Manpower Planning, with Special Reference to Thailand* (Nedlands, Australia, June 1967).

6. The Thai *Development Plan* of 1969 acknowledged that an "examination of agricultural employment is required as soon as possible." It reports that a survey on underemployment of labor is being made by the Department of Labor, but it is not yet available. Thailand, Government National Economic Development Board, *Second National Economic and Social Development Plan (1967–1971), Annual Report 1969,* p. 58.

knowledge of this fact rests the rationale for an employment policy and strategy[7] which implies:

1. More investment in human as compared to physical capital; development of skilled and semiskilled manpower; expansion of vocational education; and research and testing essential for rural development.
2. A greater volume of investment directed to rural rather than urban development, in order to reduce the gap between rural and urban incomes and slow the flow of rural workers to overcrowded urban centers.

INVESTMENT IN EDUCATION

Economic development implies change as well as growth, and it depends to a large extent upon alterations in human behavior patterns. Investment in education is a primary method of altering human behavior. This is true for general education (value-changing) of youth who will eventually become farmers and for extension-type education that can assist the farmer in adapting his "economizing behavior" to new inputs and production techniques. But even if values, attitudes, and decision making of rural youth are altered, a time lag occurs before these people can implement decisions as farmers. Development in agriculture is therefore more immediately dependent upon sound extension education than it is upon extended primary or secondary education.

7. The importance of rural employment problems is stressed in the 1969 Thai *Development Plan,* which reports that a survey conducted by the Ministry of Agriculture and the Department of Community Development showed a farmer can find himself fully employed for about 87–100 days per year on average (pp. 69–70). These working days occur during the approximately six-month rice season. Occupational training in agriculture for better efficiency and nonagricultural training for employment in off seasons are part of the Northeast Development Plan. Apart from this sort of policy it is recommended that the government "promote rural construction projects through maximum utilization of nonskilled manpower for the increase in rural employment during off seasons."

How much to insist on either or any type of education is to some extent a function of how much one knows about the return, in productivity, to be expected from it. "Manpower forecasting" in any detail is part of such an approach. To undertake such forecasts properly requires not only data on the current manpower situation but also ability to make projections of the labor supply in the aggregate, of future manpower requirements by economic sector, and of levels of education expected to be achieved by the labor force; it requires knowledge of occupational distributions and requirements, and many other details. Under these conditions, an adequate assessment of expected manpower shortages is possible only for Thailand. The Vietnam situation is in flux, but at least rough estimates have been made; Cambodia is an unknown quantity; and Laos lacks a recent census, so that sophisticated manpower projections are unavailable.

A disturbing contrast, shown in Tables 3 and 4, is the great preponderance of university students and graduates in the fields of law and social science as opposed to agriculture and engineering in Cambodia, Thailand,[8] and Vietnam. Imbalances will exist between the results of the educational system and the manpower needs of the predominantly agricultural riparian countries. The choice among fields of study reflects cultural values in the Mekong Basin, as elsewhere. As discussed in the next section, the value placed on formal knowledge by Basin peoples does not emphasize skills in manipulating material resources. A liberal arts degree may be a "key to social status and hopefully secure a prestigious berth in the civil service." [9]

Educational planning which concerns itself exclusively with matching the output of the educational system with manpower requirements can ignore the society's valuation of the total

8. Buasri Thamrong, "Some Suggested Policies for Educational Investment in Thailand," in U.S. Department of State, Agency for International Development, *Manpower in Economic and Social Growth,* Proceedings of the Sixth International Manpower Seminar, June 1-Aug. 13, 1966, p. 229.

9. William W. Lockwood, "The Rise of Educational Opportunity in Asia," *International House of Japan Bulletin,* No. 18 (Oct. 1966), p. 18.

benefits from different fields of study. However, an educational system which pays no attention to prospective manpower needs is likely to prolong the adverse effects of educational priorities inappropriate for economic growth.[10] For development goals to be met, traditional educational systems must change their goals, structure, content, and method, rather than expand along existing lines. It is difficult to see how comprehensive development planning and financing can avoid being involved in issues of this sort.

CULTURAL VALUES

The peoples in the Basin have gradually formed their values—judgments of what is significant and desirable to them in life—from their own experience and their understanding of the universe, the natural environment, and man. Their values also express their understanding of their material and spiritual opportunities in life, given their technology, social institutions, and beliefs. The current values held by peoples in the Basin are influential in shaping the material and social rewards they can expect for their choices and achievements in life. Their values are thus an important element in the inclination and capacity of both urban and rural people in the Basin to take on the risky choices and difficult achievements involved in modern agricultural production and marketing. A central premise of this analysis is that a people's value system is congruent with its material and social reward system and that this relationship is necessary for a meaningful life. Although these two systems are closely linked with many aspects of their life, people are capable of changing both systems in response to changing life conditions. People in the Basin could thus maintain a sense of meaning and also undergo extensive agricultural development *if* it occurred in a manner enabling them to modify their value and reward systems in relation to each

10. Philip H. Coombs, "The Need for a New Strategy of Educational Development," *Comparative Education Review,* Vol. XIV, No. 1 (Feb. 1970), p. 79.

TABLE 3. Third Level Education:

Country	Year	Sex	Total[a]	Humanities	Education	Fine arts
Cambodia	1956	MF	485	—	—	—
		F	35	—	—	—
	1960	MF	724	—	—	—
		F	'40	—	—	—
	1962	MF	1,529	215	217	—
	1965	MF	5,851	861	2,089	134
		F	800	138	498	14
Laos	1965	MF	146	—	—	—
		F	24	—	—	—
	1966	MF	216	—	—	—
		F	35	—	—	—
Thailand[b]	1949	MF	30,143	353	—	81
		F	2,522	308	—	8
	1959	MF	35,631	624	2,033	551
	1963	MF	42,191	806	5,027	736
		F	12,096	690	2,684	139
	1964	MF	35,953	1,430	5,504	765
		F	12,651	1,241	2,826	129
	1965[c]	MF	36,403	1,469	5,334	765
		F	12,236	1,261	2,833	149
Vietnam	1960	MF	11,761	2,869	975	106
		F	2,082	704	215	1
	1965	MF	27,105	8,221	998	431
		F	6,553	2,585	256	22
	1966	MF	31,643	7,907	1,416	924
		F	8,029	2,492	392	21

Source: UNESCO, *Statistical Yearbook*, 1968.

— Nil or negligible. ... Not available.

[a] Sum of categories for which data shown.
[b] Universities and degree granting institutions only.
[c] Decreased enrollment figures in 1965 due to adoption of entrance examination system at Thammasat University in 1960.

Distribution of Students by Field of Study

Law	Social sciences	Natural sciences	Engi-neering	Medical sciences	Agri-culture	Not specified
229	—	—	—	182	74	—
17	—	—	—	18	—	—
409	—	36	15	264	...	—
21	—	7	1	11	—	—
512	32	119	71	363	...	—
490	163	164	82	483	65	1,320
22	5	16	1	54	—	52
77	Included	—	—	69	—	—
7	with	—	—	17	—	—
103	Law	—	—	113	—	—
28		—	—	7	—	—
23,907	3,213	999	476	812	302	—
376	1,062	384	8	376	—	—
9,411	16,365	1,943	1,417	2,315	972	—
9,322	17,891	1,567	1,741	3,380	1,721	—
422	5,611	515	31	1,629	725	—
9,081	10,590	1,522	1,787	3,587	1,687	—
453	5,277	533	30	1,758	404	—
8,757	11,277	1,522	1,913	3,588	1,688	—
429	4,842	523	39	1,757	403	—
2,359	54	3,160	225	1,906	107	—
255	12	328	1	557	9	—
6,336	865	5,383	345	4,207	319	—
1,125	157	860	—	1,526	22	—
8,871	878	6,192	350	4,430	385	290
2,043	155	1,059	—	1,782	37	48

TABLE 4. Third Level Education: Distribution of Graduates

Country	Year	Sex	Total[a]	Humanities	Education	Fine arts
Thailand	1964	MF	5,627	168	1,078	104
	1965	MF	5,528	203	913	223
		F	2,194	168	376	34
Vietnam	1957	MF	273	12	65	3
		F	59	—	22	—
	1960	MF	521	32	177	4
		F	112	7	53	—
	1965	MF	983	184	52	8
		F	220	36	6	—
	1966	MF	1,392	240	297	7
		F	346	53	70	—

Source: UNESCO, *Statistical Yearbook*, 1968.

— Nil or negligible. ... Not available.

[a] Sum of categories for which data shown.

other. Change of such scope, however, must involve entire nations, not simply the village farmers.

Development programs often begin with the introduction of new special economic activities or of new particular social institutions. While these are likely eventually to become a part of the people's value structure, the changes involved are normally uneven, reflecting uneven adjustments among the aspects of a people's experience. In short, development tends to be disruptive in traditional societies; but the greater the extent to which novel economic opportunities and activities can be related to the institutions, values, and rewards of a people, the more smoothly development can become a part of their life. Planned development, by the Basin governments and outside agencies, can work out manageable accommodations between the economic and social requirements of effective development and the requirements of continuity and a meaningful life for the people concerned, if the accommodations are specifically sought in both directions.

by Field of Study, Thailand and Vietnam

Law	Social sciences	Natural sciences	Engi-neering	Medical sciences	Agri-culture	Not specified
1,127	1,922	165	317	518	228	—
765	1,202	357	384	955	526	—
401	456	135	8	492	124	—
70	...	11	23	89	...	—
11	...	2	1	23	...	—
119	...	21	63	94	...	—
14	...	6	1	31	...	—
305	...	98	86	250	...	—
70	...	16	—	92	...	—
288	...	127	61	372	...	—
54	...	22	—	147	...	—

The basically similar values achieved by the Basin peoples emerged in the course of a long-stable adaptation to wet rice culture in a monsoon climate, within the fold of the Hindu/Buddhist tradition in the upper part of the Basin and within a compound Hindu/Buddhist/Confucianist/Laoist tradition in Vietnam. As regards the implications of these two traditions for agricultural modernization—the main center of our interest—their differences are less important than similarities in both areas.

The monsoon environment was a difficult one in which to achieve wet rice agriculture. In many parts of the Basin, rainfall has never been sufficient for adequate rice cultivation, forcing farmers to manipulate their simple technology to produce a subsistence crop. The environment has offered few alternatives, especially in the long dry season, to the traditional simple technology available to farmers in the Basin. Within their understanding of nature and their limited resources, the farmers are thus quite pragmatic in their efforts to make a living, generally open to novel alternatives with conspicuous, high short-term return and low apparent risks, and generally confident of their ability to make a modest living in their own environment.

With an eye to their relation to agricultural development, we shall examine value positions of farmers and officials regarding wealth, security, prestige, hierarchy, local solidarity, virtue, and the organic nature of the world, keeping in mind that differences between people in the Mekong Basin and other peoples lie in specific goals and ways of attaining them rather than in general categories.

Wealth. Urban and rural people in the Basin clearly define wealth as a desirable good and want more of it. Almost everyone in the Mekong Basin has been touched by the rising flood of consumer goods, by a rising level of expectations for such goods, and by an even faster rising flood of inflation. Villagers have made great efforts in agriculture and in urban wage labor to acquire wealth. As indicated below, however, both villagers and townsmen seek numerous social rewards in life as well as wealth. The importance of these social rewards has led some officials and foreigners to the mistaken conclusion that villagers are bound up in "tradition," which inhibits them from responding to modern economic incentives. These social rewards support *and* contradict acquiring more wealth, but greater wealth remains nonetheless a major positive value. Several studies of village society and agriculture in the Basin[11] have affirmed that farmers do respond to new economic opportunities, as mentioned in Chapter 3, within their realistic economic capacities and within their own understanding of the likely risks and profitability.

Although their material desires are increasing, peasant farmers in the Basin are still realistically modest in their expectations of

11. See, for example: J. R. Behrman, *Supply Response in Underdeveloped Agriculture* (Amsterdam, 1968); J. Ingersoll, "Human Dimensions of Mekong River Basin Development: The Nam Pong Project in Northeastern Thailand" (mimeographic report to U.S. Agency for International Development, Bangkok, 1968); C. Keyes, "Peasant and Nation: A Thai-Lao Village in a Thai State" (dissertation, Ithaca, 1966); M. Moerman, *Agricultural Change and Peasant Choice in a Thai Village* (Berkeley and Los Angeles, 1968).

their future economic opportunities. Long experienced with crop failures, recently experienced with price fluctuations, and not yet directly affected by expanding economies, these village farmers do not think, as do development professionals, in terms of expanding horizons. Mekong villagers share the widespread peasant view of the "limited good": as one villager acquires more of the good things of life (land, wealth, social standing, influence), the total supply is diminished for others.[12] Failure to relate this rather static view to peasants' actual experience in their natural and social environment has misled some outside observers to conclude, for example, that Lao farmers will simply work half as hard if they get twice as much yield. Values of Mekong farmers and townsmen can only be interpreted in the context of *their* life conditions, including their definitions of their own opportunities in life.

The central ambition of government officials used to be the attainment of more senior rank and higher royal honors; these could be relied upon as sufficient to provide them either with enough wealth as perquisites or with opportunities to accumulate wealth by sanctioned use of their official position. Contemporary officials living in large urban centers compare their lot with that of much wealthier businessmen or foreigners, and find their official income ever less adequate, especially as their ambitions for senior positions imply a relatively high standard of living which the widening gap between fixed salaries and constantly rising prices makes it ever harder to afford. One solution, graft, which tends generally to increase with seniority, could at one time be carried on with relative legitimacy. But today, it encounters an atmosphere of increasing public criticism.

Unlike farmers, government officials can look forward to a promotion ladder with successive though modest increments in salary and perhaps substantial improvement for unofficial gain as their careers progress.

12. For relevant observations made in a recent study of Mexican village life, see G. Foster, *Tzintzuntzan: Mexican Peasants in a Changing World* (Boston, 1967), Ch. VI.

Security. The desire of Basin farmers for more wealth is principally a search for security and prestige, that is, enough land to produce adequate crops for subsistence with a surplus for cash sale, and enough land to pass on to their children. For peasant farmers who have recently entered commercial agriculture, wealth in the form of land is the most important key to security. The family household is the unit that controls land, and the household head normally makes decisions regarding cultivation. Land tenure patterns are thus a major element of rural security. They have been described in Chapter 3 above.

In addition to their striving to acquire adequate land in order to achieve economic security, Basin farmers also attach great importance to achieving social security within family units and within groupings of *mutual support.* Veneration of ancestors and honoring of elders is probably strongest in Vietnam, but peoples throughout the Mekong Basin place a strong value on recognizing their obligations and debts to their parents, especially in the latter's old age. Reciprocal help is essential at peak seasons of rice cultivation—the times of transplanting, harvesting, and threshing. A strong ethic of reciprocal help—far beyond agriculture—exists throughout Southeast Asia. Through a continuing stream of exchanges of goods and services with selected relatives and friends, village farmers manage not only their rice cultivation but their social existence. Most of these mutual assistance groupings are very small, are without formal organization, and may be broken up at any time. Although Basin villagers uniformly prize village harmony and unity, most of their relations with each other are carried on within small groups of kin and friends, the major structure of which is provided by the personality of the dominant member.

Prestige. A major social goal for Basin peoples, rural and urban, is prestige, based on wealth and on relations with esteemed senior people. Individual achievement is a lesser consideration. Elaborate family life-cycle celebrations, temple festivals, and the publicity normally attending people's contributions to these occasions all are part of a prestige-conferring pattern of conspicuous distribution of goods within village society. With the coming of many commercial goods and services from urban centers, villagers have been shifting toward a pattern of modern conspicuous consumption mixed with traditional conspicuous distribution. Although appreciated for their convenience, many of the new items and styles are also valued as public indications that villagers are keeping up with the new, urban way of living.

As people in the Basin acquire prestige from, and accord it to, each other, they tend to make judgments of people's worth based on status-centered and person-centered considerations more than on those of individual achievement. A member of a prominent family can be highly regarded whether or not he has himself contributed greatly to the family's position. Respect has long been naturally accorded to the well born, who, in the karmic moral order, must have deserved any social benefits which they receive in their current life.

Among the social elite, as among villagers, people gain respect and prestige largely on the basis of their relationships with people of high status. Civil service promotions in all four Mekong countries are still influenced significantly by the types of personal relations one maintains with highly placed officials. Young officials ambitious for promotion prefer a job in the Ministry close to the sources of high power and prestige to an assignment in a remote rural place.[13] Formal education has become increasingly important as a qualification for young people in obtaining good government positions, but the traditional person- and status-centered values still influence promotion.

Hierarchy. As the Vietnamese adopted and modified Chinese views, and as the other peoples of the Basin adopted and modified traditional Indian views, kingdoms grew up throughout the Basin which achieved a thoroughly hierarchical view of the universe and

13. Many foreign advisers and technicians who have complained about the long and frequent absences of their national counterparts from rural projects have not appreciated how important it was to the latter's promotion to appear as frequently as they could in the presence of their superiors.

human society. All creatures, from the humblest animals to man to the deities, were placed on rungs of a great ladder of life. Although kings no longer rule at the apex of national hierarchies in the Mekong world, widespread respect for royal hierarchical social tradition remains. Interpersonal relationships are formed in terms of status seniors and status juniors. For most urban and rural people in the Basin such relations are still more comfortable and natural than egalitarian ones. This preference has far reaching implications for the functioning of development institutions designed to accomplish wide dissemination of innovations, and pertinent feedbacks up and down the hierarchy and among equal levels of authority. For example, senior people in official hierarchies would not feel free to delegate their authority or to distribute their decision making powers very extensively to their staffs or to field subordinates, nor would they be rewarded for it. Neither the senior nor subordinate officials would expect the latter to take much action or initiative without authorization from the capital.

Granted inevitable delays and distortion of messages passing through many layers, decisions and information within most official hierarchies in the Basin have moved downward from Cabinet to village level with great effectiveness; but apart from authorized reports to the top, feedback or corrective types of information are much less complete or effective. The same is true for communication from a given level in a hierarchy to a comparable level in another hierarchy, such as another ministry. Emphasis on authority makes it very unrewarding for a person or an organization to invest much effort in extending information horizontally to, or coordinating with, other comparable organizations. Related organizations must rely upon a common superior to authorize coordination between them. Hence, coordination is particularly complicated and weak in Basin programs. These difficulties constitute important reasons for inventiveness in strengthening voluntary associations at the village level and beyond as well as private enterprise, despite the delicate issues that, in the latter case, arise from ethnic rivalries.

When functioning well, social hierarchies have been effective in securing the well-being and looking after the interests of members at different levels. The function of the superior—whether official, friend, relative, neighbor, or priest—has been to preside, to lead the others, and also to care for the welfare of the subordinates. The function of the subordinate has been to follow, to provide loyalty, obedience, and praise to the superior who acts as patron. Despite the vertical emphasis of social relations, people have been free to move about in all four Basin countries and to form new relationships with new patrons.

Local Solidarity. The degree to which inhabitants of Basin villages have identified with, have felt solidarity with or loyalty for their communities is a difficult issue. Yet it is an important one in determining appropriate units for organizing planned development. Most Basin villages are small enough and homogeneous enough that villagers generally share a common cultural tradition and commonly associate with each other. But feelings of common identity, solidarity, and loyalty within villages are by no means complete.

Most observers of village community life in the Mekong countries have pointed out aspects of both community solidarity and segmentation. The inhabitants of most Basin villages share a common culture; feel solidarity when compared with or pitted against the outside world; feel more comfortable in each other's company and less constrained to be on their best behavior. But villagers also commonly have smaller groups with which they identify within village life, such as neighborhood groups and kin groups (lineage groups in Vietnam).

Traditional inclinations of villagers in Southeast Asia to avoid entanglement with the authorities of the larger society are succinctly stated in the Vietnamese proverb: "the word of the Emperor stops at the village gate." Nonetheless, most villagers now see their village as their residence but no longer as their total community, as they slowly become participants in their larger political and social units. The social, economic, and cultural gap which still separates village and urban life, however, is a major fact in these nations, as well as a major impediment to the capacity of the four governments to carry out their policies of national development and modernization.

Virtue. Although the peoples in the Basin are devoutly attached to their religion, they are by no means ascetics. Most of the objectives of their moral restraint and virtuous behavior are emphatically the values of everyday living in this world. Apart from vegetarian practices by some Vietnamese Buddhists and a strong reluctance among most Buddhists in the Basin to kill any major form of life, the Buddhism of everyday life is not other-worldly and not at all contrary to the kind of behavior entailed in agricultural development.

The various Basin peoples have a highly moral view of human conduct derived from the doctrine of karma: all one's fortunes and misfortunes are his own responsibility since they are the outcomes of his own good and bad choices during the current life or a previous life. In these people's view, by properly guiding his own conduct, a person can enhance his chances for the good things of life. In addition to the choices for good or ill a person can make, he can call upon several different types of divinatory and magical specialists to ascertain or improve his fortunes. Most people in the region thus probably feel that they have more elaborate and more effective spiritual means to influence the course of events than technological means to manipulate their lives or their environment. Thus, virtue is both a moral goal and a means by which people in the Basin can aspire to improve their condition. Neither a remote ideal nor a constant attainment, meritorious conduct is a useful ingredient in bettering one's position, materially as well as spiritually.

Organic Nature of the World. Although now intermixed with modern scientific knowledge and beliefs, the traditional view that the world is an organism rather than a mechanism and that mystical forces relate human life to the cosmos remains as an active influence in Basin villages and urban centers. For example, farmers continue to observe an astrologically auspicious time for beginning first plowing. The Thai king and government establish this time by conducting an official cermony in Bangkok. More generally, Southeast Asian peoples use astrology to guide the proper timing of events, and geomancy to orient buildings properly in space. Men still seek sacred knowledge for spiritual as well as magical purposes and do so by the cultivation of moral restraint, disciplined meditation, and prodigious memorizing.

Secular experience like that in agriculture, on the other hand, has been conventionally acquired by sustained exposure. Neither explicit observation nor explicit teaching has seemed necessary to farmers. Although the technology of wet rice cultivation is a highly evolved tradition, its practitioners are aware simply that they carry it out as the old people did before them. Thus agriculture does not in their minds require formal study, scientific research, or special training. They have merely kept some of their best seeds each year and planted them the following year to cultivate them in the conventional manner.

Consistent with the traditional organic view of the world has been a view that the purpose of knowledge "is to clarify, explicate, and elaborate upon truths already made known by ancient seers. This approach stimulates research whose limits are predetermined, whose function is conservative and resistant to novelty.[14] This view of truth as an entity rather than a goal is inimical to the spirit of modern scientific inquiry and technological innovation, on which western development has so heavily depended.

Although particular aspects of scientific thought, and a view of modern science as fashionable, are widely spread throughout the Basin, the traditional view of formal knowledge as an established substantive entity is still influential. Thus students may have thoroughly memorized the multiplication tables and later the atomic weight chart without being able to apply either one to the solution of problems. The findings of scientific experiments become part of the established substantive learning, but scientific research is not yet widely understood as a basic method for discovering natural reality. In this substantive view of knowl-

14. W. Bradley, "The American Research Effort on Southeast Asian Development Problems," *Asia* (1968), p. 10.

edge, the extremely precise measuring of units of time, space, or weight does not seem very important. Thus, officials of an agricultural experiment station as well as the farmers they are presumably trying to influence do not all appreciate the critical importance of precise timing or measuring of water or height of rice plant in the irrigated cultivation of the high-yield rice varieties. The point here is not that riparian nationals cannot or have not become very good scientists; they can and they have. The point is that the pervasive cultural view of knowledge does not set the stage for effective rewards for most of the people, including many of those hired at research and demonstration centers, to be committed to the demanding and elaborate procedures of precise scientific observation or controlled experiments.

A drastically higher rice yield does, of course, represent a very practical reward for the grower, but it will take time for its appearance to be associated as a matter of course with scientific farming rather than with more deeply rooted cultural and religious views. Moreover, pragmatism will not fill the bill by itself.

> Few people and very few cultures in Asia are able to live among the happenings of our day-to-day life without some sense of meaning—a sense which a transcendental vantage point provides. Man's mortality, the cycle of birth and death, growth and decay, the seeming senselessness of much of human experience only becomes bearable within the context of some kind of eternal truth and reality. And especially in Asia, where religions have not only been roads to the salvation of the individual soul, but also have helped shape systems of social organizations. This aspect should be taken into account in any analysis of social dynamics.[15]

SOCIAL INSTITUTIONS

It is largely through their participation in social institutions that people formulate, express, and modify their cultural values.

15. "Religions and the Development Process in Asia," paper presented by H. E. Soedjatmoko, Ambassador of Indonesia to the United States, at the Asian Ecumenical Conference for Development, Tokyo, July 15, 1970.

Thus, development programs in the Basin will be self-sustaining only as they become rooted in, and help effect changes in, some of the major village, government, and commercial institutions. Development entails the perfection and spread of technological, economic, and organizational capabilities among more people; the increase in these new skills results in the differentiation of existing social institutions into more autonomous units with more specialized functions. With continuing differentiation, social order and even further development become difficult unless the segmented institutions become somewhat reintegrated in a new order which reflects the more modern bases of power and values that have been taking shape.[16] What follows is no more than a glimpse—albeit a necessary one—of the rural social structure[17] of which Mekong development programs must become a part and which those programs must help modify. While not every one of the elements described carries a "lesson" for development, immediately leading to a "strategy," we feel strongly that a lack of understanding of these traditional institutions can and does impede the designing of improved, modern ones.

In South Vietnam the paternal lineage or a segment of it is an important unit of identity, but in all four countries the *family household* remains the basic institution of village life. Members of the family household manage their agricultural holdings, try to acquire additional holdings, and pass them on to successive generations. The traditional Vietnamese pattern of inheritance through the son, particularly the eldest, has been substantially modified and now frequently includes daughters. Cambodian families tend to distribute their land to both sons and daughters, while the northeastern Thai and Lao families tend generally to favor daughters with inheritance.

16. For example, the numerous functions of the formerly self-sufficient peasant family have become differentiated into organized temple, school, and headmen, all of which now form part of a more sophisticated village social order.

17. For a discussion of commercial and governmental institutions see Ch. 3.

A second village institution of loose organization but great importance is the *mutual help network*. Each farmer tends to establish a network of relatives, neighbors, and friends within which he exchanges labor in the course of the rice cycle. It is important for village organization that the network of each farmer is a distinct and different grouping, as are the networks of relatives of each individual. A village is not composed, for example, of ten or fifteen regular mutual help groups, but rather the network for every single farmer is a somewhat distinct one.

The most elaborate village institution in all four countries is the *temple*. Although a priest genuinely concerned with village development can be the single most effective local advocate of development, the general body of villagers attending the temple and the common but universal temple committee of elders have not thus far generally lent themselves to becoming village institutions for planned development.

Until recent generations the running of the village *school* was an additional activity of village priests. Young boys received basic instruction in literacy and in some doctrines of Buddhism. In this century schools have become differentiated from the temple and manned by paid civil servants. No longer concerned with transmitting sacred knowledge and doctrine, village schools now transmit elementary literacy but do not yet prepare village children either for more effective modern farm life or for employment in urban life. The degree of parental interest in village schools and of pupils' diligence tends usually to be directly related to the degree of likelihood that the village elementary school can prepare children for secondary school in town.

Finally, there are the village *headmen*, and in some parts of the Basin, the *village council*. These traditional institutions have become increasingly dependent upon modern central government power.

A fairly recent arrival in most Mekong villages is small scale commerce—a general store, a coffee shop, a rice mill, and a bus to town. These rural commercial institutions, more numerous in Vietnam and Thailand than in Cambodia or Laos, are owned partly by ethnic nationals and partly by resident aliens, such as the Chinese in all four countries and the Vietnamese in Cambodia and Laos. Operated with very limited capital and on a very modest scale, they are generally owned by wealthier villagers, rather than by townsmen extending their operations. These village businesses are rapidly becoming more important in village commercial life as farmers increase their participation in the cash economy. Farmers are now sending much more of their rice than previously for milling at village rice mills instead of hand milling it at home.

SUMMARY

Several general implications emerge from this review of Basin human resources. First, people in the Mekong Basin are generally interested in acquiring more wealth, but they remain at their relatively low level of development for a number of complex and closely interrelated reasons. Their physical resources, abilities, and rewards are still organized around their traditional existence more thoroughly than around their modern attempts to improve their lives through development programs. The hierarchical, centralized orientation of government agencies—upward toward the capital more than downward and outward toward the people—is still more consistent with the older function of the government elite to maintain order and harmony with the universe; the simple, general, informal institutions of village life are still more consistent with the former condition of rather isolated, self-sufficient villages remote from the seat of government; and the commercial institutions remain largely as ethnic enclaves in the larger society and economy, whose members are disliked and mistrusted and whose close knit life reflects their own incomplete integration in, or identification with, their larger society. Much better organized and economically stronger than farmers and thus able to control prices paid to and by farmers, these firms have, on the other hand, been obliged to make heavy graft payments to officials in order to compete with some government corporations and with a few international firms.

These institutional features are the social concomitant of low levels of wealth and production, limited technology and trained manpower, and relatively low value placed on productivity or high production as an end in itself. It would be unrealistic to anticipate, and naive to plan, substantial improvements in either physical resources or human resources for agriculture without providing rewards for changes in current values and institutions.

A second general implication is, therefore, that the peoples and governments in the Mekong Basin can achieve substantial development only by further modifying some of their social and cultural organizations. Such extensive changes in conduct will require increases in their technical knowledge and ability and considerable change in their system of rewards. These changes will be necessary not because their current social and cultural organization is inferior but because such changes are a part of substantial growth in the level of economic performance and achievement in any society. As emphasized above in the discussion of values, these changes can be successful only if they are undertaken as *mutual* accommodations between the current values, institutions, and rewards of Basin countries and the requirements of Basin agricultural development, with the active participation of Basin villagers, merchants, and officials.

A third general implication is that outside agencies will have a very difficult and sensitive role to perform in assisting Mekong governments and peoples to become aware of, and then to make, the necessary social modifications. The large number of inadequately used or malfunctioning irrigation projects already in the Basin provide ample evidence that development does not flow from erection of physical structures unless they are accompanied by administrative, economic, social, and cultural change. How inputs of funds and foreign personnel can help is far from clear. Providing assistance in sensitive areas of social change also creates opportunities for cultural arrogance. Against this, one must consider the consequences that are obtained when outside agencies provide assistance for physical facilities only and reap results which disappoint the recipients.

If international development agencies have no choice then but that of helping those governments to make their societies more capable of undertaking vigorous development, how do they go about it? Are there ways other than encouraging the widest possible study and discussion of such needed social changes among officials, researchers, and peoples of the Basin? We consider this matter in the next chapter.

5

A Strategy for Agricultural Development

INTRODUCTION

The previous two chapters have shown that the present physical and institutional infrastructure serving farmers in the region is incapable of supporting the transition to widespread employment of modern production techniques. This chapter attempts to sketch in broad strokes the measures which should be taken to remove the barriers to improving the infrastructure. The emphasis is on the sorts of investments and institutional changes needed, rather than on details concerning specific investments and programs.

We have made no attempt to estimate the quantities of fertilizers, pesticides, improved seeds, farm machinery, and other inputs which would be required to effect the transition from primitive to modern production techniques, nor have we considered the extent to which these inputs should be imported or produced in the region. These are questions of tactics rather than strategy. We believe, however, that if imports of these inputs are freely permitted at world prices, there is no reason why they cannot be made available to Basin farmers in the amounts needed on economically attractive terms. Should supply limitations emerge or prices be too high, the problem almost surely will be failure to remove bottlenecks in the institutional and physical infrastructure, or in government import policies.

INVESTMENTS IN PHYSICAL WORKS

The most pressing needs are for investments in water management systems, in the transport network, and in storage and processing facilities. With respect to investments in transport, storage, and processing facilities, we are not prepared to go beyond the obvious statement that where the inadequacies of such facilities at present inhibit the expansion of agricultural production, the facilities must be improved. Questions of the scale and precise nature of such investments, and of their specific location in the region, involve details which are outside the bounds of this report.

Some useful general statements can be made, however. There is a strong argument for making these investments in projects to improve the operation of *existing* works rather than in construction of *new* large scale projects. The argument rests in the fact that present water management systems employed in the region are grossly inefficient, particularly at the farm level. Until these systems are improved, the return to investments in large new projects is likely to be low. Well chosen investments designed to increase the efficiency of existing systems, however, may yield very high returns.

The general nature of the investments needed in water projects is well described in the Asian Development Bank's *Asian Agricultural Survey* (1969), which asserts, with reference to Asia generally, that

> priority should be given to a program of intensification of the use of present water resources and investment in early return projects. That is, the immediate future emphasis should be on making better use of existing irrigation facilities and developing minor irrigation projects. And in planning future irrigation, greater attention must be given to water distribution to farm fields and to water removal through drainage.

In another part of the same report it is stated that

> present irrigation and drainage projects, both in operation and under construction in practically every country in the region [i.e. Asia] need further development *in terms of investment on ancillary items such as storage tanks, secondary and tertiary canals, etc.* In many cases even immediate improvement in the maintenance and particularly the manage-

> ment is urgently required in order to realize the fullest possible potential. [Emphasis added.]

Most farms in the region lack the physical works needed to assure controlled application of water to all parts of the fields. The *Asian Agricultural Survey* observes, again with respect to Asia generally, that

> one of the outstanding characteristics of the irrigation facilities in this region is the lack of a terminal network to spread water over the farm fields. Water is supplied to individual paddy fields simply by overflowing from an upper to a lower lying field. . . . The provision of a terminal water distribution system, based on the consolidation of fragmented farm lands and combined with a network of farm roads, is basic to the improvement of rice culture.

The creation of such a system may not be easy to accomplish, but whether it is basic or not, it surely would be immensely helpful.

In a paper given at the Regional Seminar on Agriculture, David Hopper, in dealing with the problem of water control at the farm level, noted that the predominant irrigation pattern in Asia is to let the water flow "from inlet channel to drainage outlet, field to field down and across a terraced slope." While this procedure may be satisfactory for traditional agriculture it is not good enough for modern intensive farming.

> Control at each field commanded by a major water system depends on the construction of a terminal network of interlocking supply channels and drains serving fields that have been levelled and shaped to permit a uniform water distribution to all parts. Such networks are not cheap to build. But water control is the major barrier that Asian agriculture must hurdle if it is to fulfill its promise of abundance.[1]

Investments to improve water distribution systems at the farm level clearly are of major importance. Improvements are needed

1. *Papers and Proceedings,* p. 32.

also, however, in systems for bringing water to the farm gate. As noted above, we believe that the main emphasis should be given to the development of many relatively small scale projects rather than of a few very large ones. The sorts of projects we have in mind are illustrated by a program undertaken some years ago in Thailand to build small local reservoirs, usually referred to as "tanks." How many of the 1,000 tanks originally planned have been built to date is not clear. In 1969 there were about 160 tanks in existence, most of which still did not have water distribution facilities. A report by the SCS–PASA[2] teams contains a description of 6 such tanks. These range in capacity from 1,700 acre-feet (2,000,000 m^3+) to 4,900 acre-feet (6,000,000 m^3+), with an average of 3,050 acre-feet. The area reported as "irrigated" ranges from 520 acres to 3,040 acres and averages 1,850 acres. This works out to a water supply or water "use" varying from 1.16 acre-feet per acre to 5.6 acre-feet per acre, with an average of 1.65. It seems more likely that "irrigated" area refers to the area to which water *could* be physically delivered and that some of this area would not be irrigated in any single year.

Another uncertainty is that we do not know how representative these 6 tanks are of the 160 or so that have been built to date—or of the 1,000 eventually to be constructed. However, assuming that they are representative, then the 160 tanks would have a capacity of nearly 500,000 acre-feet and an associated irrigable area of nearly 300,000 acres. Even if the latter figure is discounted to, say, 100,000 acres that could be irrigated during the dry season, this is still a significant amount. If all 1,000 tanks originally projected were to be built, the expansion in irrigated area would be very large indeed, and because of the opportunities created for double cropping the contribution to production would be even greater.

Aside from their production potential, these tank projects also provide an opportunity for water management studies, demonstration, and operation on a pilot plant basis. Here, on a relatively small and hence more manageable scale, and under such condi-

2. Soil Conservation Service–Participating Agency Service Agreement.

tions that many farmers could observe the results, the best water management practices that the technicians of the region or visiting technicians could devise could be developed, tested, and used. If water management cannot be made to pay and to attract farmers to its use on this scale, then its prospects in larger scale projects are indeed dim.

Another example of water management projects which could be developed on a relatively small scale is provided by the polders which may be built in the Mekong Delta in Vietnam. These would consist of some thousands of hectares each. Each polder would be surrounded by its own earthen dikes to provide flooding from streams; each would have some means of pumping out excess water; and each would have some means (usually the same pumps) for pumping irrigation water from the stream or canal into the polder during the dry season.

This method of water management is possible only in areas sufficiently far down the Delta that the flood depths are not too great, and only in areas sufficiently far up the Delta that salt water intrusion from the sea is not a factor. These conditions are met in a relatively small part of the entire Delta; some areas below Vam Cong and above Can Tho will qualify. Moreover, for at least two reasons the extent of this type of development cannot be very large until some mainstem dams on the Mekong provide a larger dry season flow. First, if an attempt were made to take much water out of the canals and rivers, the salt balance would be upset in other areas, perhaps causing damage as great as any values created. Second, too much diking would reduce the capacity of the Delta to absorb floods, and thus poldering must wait until mainstem dams reduce flood hazards. Thus qualified, however, the polder projects hold great promise, for the soils concerned are among the best in the entire Basin. Under these circumstances, improvements in water management set the stage for intensive and highly productive farming.

Last but not least are the water development projects included in the Amplified Basin Plan for 1971–80 (excluding those that are power projects only). The list of projects included by the Mekong

Committee in its short range plan (as of June 1970) is shown in Table 5. There are two projects in Cambodia, the Prek Thnot being under construction and the Battambang approved for construction with some initial work begun. The Nam Ngum project in Laos is well under way. Two projects, the Nam Pong and the Nam Pung, in Thailand are listed as "completed," but presumably this applies to the dam and power houses only; neither is yet in a position to supply water to the full area listed

TABLE 5. Water Development Projects Included in the Amplified Basin Plan for 1971–1980

Project	Area scheduled for irrigation (1,000 ha.)	Status 1970
Cambodia:		
Prek Thnot	35	Under construction
Battambang	68	Some initial work
Laos:		
Nam Ngum	35	Under construction
Thailand:		
Nam Pong	53	Completed
Nam Pung	14	Completed
Lam Dom Noi	24	Under construction
Lam Pao	62	Under construction
Lam Pla Plerng	13	Under construction
Nam Takong	34	Under construction
Nam Oon	32	Under construction
Vietnam:		
Polders in the Delta	40	Planning
Upper Se San	22	Planning
Upper Sre Pok	24	Planning
Total	456	

Note: Excluding power aspects of these projects and excluding power projects.

as "irrigated," and actual irrigation water use to date for each is nil. Five other projects in Thailand are listed as under construction, as well as three in Vietnam in the planning stage.

Development of these projects depends upon the absence of extreme disorder and violence in their respective areas; the situation for the various projects is highly variable at the present time, and there are no means of knowing what the situation might be between now and 1980. Under the best of circumstances, the areas listed as "irrigated" could not be fully productive by 1980. Some of the projects are listed in the Amplified Basin Plan Report as "commissioned" after 1975; perhaps construction of irrigation works would not be completed by 1980. Delay between the date on which the engineers say water will be available and the date on which farmers have the irrigated land in full production is an accepted feature of projects around the globe. In addition to other delays, the necessary land leveling and establishment of crop rotations take some time. A reasonable target would be for half of the full productivity of these "irrigated" lands to be achieved by 1980; in practice, a far smaller achievement is possible, not excluding, from experience to date, the possibility that actual output from these projects may be near zero.

Even if only half the full productive potential of these "irrigation" projects is achieved by 1980, and if the full complement of other inputs (fertilizer, seeds, etc.) is applied in these areas, the achievable output is still large: 228,000 hectares, or well over half a million acres. If all the "irrigated" land were used for rice, this should increase the average yield in the wet season, by providing water when the monsoon was late or during dry spells in the monsoon season; if all this land were used for rice in the dry season, its paddy rice production could reach 1½ million tons, or over 10 percent of present rice production within the Basin. If the water were used to produce other crops, presumably the value of its contribution would be equal or greater. On this basis of calculation, the other half of the productive potential of these projects could be attained in the years immediately after 1980.

LOCATION OF INVESTMENTS IN THE REGION

If all parts of the region are to share more or less evenly the fruits of technological advance in agriculture, then the pattern of investments will have to be shaped to serve all parts of the region. Since the problem of water control is found throughout the Basin, investments to improve water management should be widely dispersed. Such improvements are particularly important in areas devoted to rice. They are essential not only for successful cultivation of the new rice varieties, but also to move the rice economy toward the system of double cropping, either with two rice crops per year or with one rice crop alternated with vegetables or some other nonrice crop.

There is some ground for believing that rice production in the future may tend to be more concentrated in Vietnam and less so in Thailand than has been the case so far. The Development and Resources Corporation report for Vietnam[3] contains projections of hectares in various crops in 1980 and 1990. These show that even by the latter date some 75–80 percent of cropland in the delta would be in rice, compared with perhaps 90 percent at present. The projected percentage increase in nonrice cropland is quite considerable, and the resulting production would both improve the diets and raise the income of delta farmers above the levels achievable if no move toward diversification occurs. Nonetheless, rice would continue to dominate completely the agricultural economy of the region. It seems a safe inference that this will be true also in the Cambodian delta.

The prospects for rice in northeast Thailand, on the other hand, appear much less favorable. According to the *Asian Agricultural Survey*, for example, the northeast has a large arable area not suitable for rice or other plantation crops because of the poor condition of the soil. A recent FAO report[4] cites a study by Dr.

3. *Projected Agricultural Production: 20-Year Mekong Delta Development Program* (New York, 1969).

4. *Recent Trends and Patterns in Rice Trade and Possible Lines of Action.*

R. Barker of the International Rice Research Institute which also suggests that the competitive position of Thailand in rice production may be weaker (or less strong) in the future than it has been in the past. Dr. Barker compares Vietnam and Thailand according to five important features of rice production with regard to the chances for spreading the new varieties. The results are shown in Table 6.

Emphasis on rice in Thailand's Northeast, even with double cropping under irrigated conditions, would not represent a good use of a costly irrigation program. "Both the terrain and the soils of the Northeast are indeed more suited for non-paddy crops and pasture," states a recent study.[5] "Insistence on producing as much rice as possible is the worst strategy that could be chosen," it concludes. As for Laos, it might be expected to meet its own rice demand, but any exports would require the consent of Thailand, over whose territory the rice would have to move. In any event, Laos would find the establishment of export markets difficult.

The relatively poor prospect for expanding production of the new rice varieties in northeast Thailand has brought other possibilities to the fore, particularly that of livestock. But promise in this field is dimmed by the lack of feed in adequate volume and quality. High temperature and humidity make for fast growth of pasture but poor quality, even when not overgrazed, as good

TABLE 6. Important Features of Rice Production, Vietnam and Thailand

	Vietnam	Thailand
Water control	Average	*Poor*
Availability of inputs	Average	Average
Yield advantage of improved over existing varieties	*High*	Medium
Disease resistance	Average	Average
Quality acceptability of new rice grain	Average	*Poor*

Note: Italics give added emphasis.

5. Ronald C. Y. Ng, "Some Land-Use Problems of North-east Thailand," in *Modern Asian Studies,* Vol. 4, pt. 1 (Jan. 1970).

pasture would tend to become. Mixed feed is expensive. There are suggested solutions, ranging from the establishment of mixed farming with cultivated pasture as a component to the importation of a Brazilian cactus that is said to serve as a satisfactory cattle feed. We are not in a position to evaluate the merits of a particular proposal but concur in the view that lack of feed is a major, if not *the* major obstacle to expansion of the livestock industry. In fact, this seems to be the case not only for cattle but also for other types of livestock including poultry, and not only in northeast Thailand but throughout the region.

The observations above may at first glance appear to fly in the face of the fact that about half of Thailand's buffalo and cattle population and over one-third of the country's pigs are found in the Northeast. Apparently, then, there is enough feed to maintain them. A brief comment is in order. First, the number of cattle has declined by over 20 percent between 1953–55 and 1965–67. The hardier buffalo, mainly a draft animal, has increased by 34 percent, or 2.5 percent per year, during the same period. Second, there is a difference between maintaining an animal, and raising it for profit as a source of meat (or milk). The weights of fully matured cattle reported from the area (500 to 650 pounds) suggest that the feed available, in association with breed characteristics and less than good health status, suffices to maintain the animal but not to make it grow to a weight that turns it into a profitable item of production.

Feed, varying with the season, consists of rice field or roadside weeds—a nutritionally poor feed; rice straw at the peak of the dry season; and bushes, grass, etc., browsed in forest areas. Lacking are cultivated pasture and specially grown feed crops. Under these circumstances most of such feed as the animal obtains is used to maintain it alive and in tolerable flesh, with little surplus for growth and reproduction.

Thus, the fact that nearly half of Thailand's cattle live in the Northeast does not contradict the urgent need for substantial improvement in the supply of feed. The consequence of such improvement would be more numerous, better fed, heavier animals fetching a higher price. Pasture cultivation might, in addition, fill an important role in crop rotation.

The stress given here to the Vietnam delta and northeast Thailand of course does not mean that fruitful investments cannot be made in Cambodia and Laos. There are possible subbasin projects in both countries as well as needed improvements in existing water management systems, in the transport network, in storage and processing facilities, and so on. Indeed, the investment strategy suggested here implies that investments would be widely dispersed throughout the Basin in comparison to the distribution implied by a strategy of mainstem development.

INSTITUTIONAL CHANGES

Turning to weaknesses in the institutional infrastructure, there appear to be three general types of services which are in particular need of attention: (1) the provision of credit; (2) the provision of knowledge and information needed to make effective use of more advanced farming techniques; and (3) the provision of health services needed to increase the vigor and work capacity of the farm labor force.

Credit. The adoption of modern farming techniques will require increased amounts of credit, which existing credit institutions are ill adapted to supply. Private banks have little experience in dealing with small farmers, nor are market forces likely to induce them to do more. Government agencies, acting directly or indirectly, almost certainly will be necessary. There is plentiful literature on the problems of providing credit to small farmers, and not a little actual experience from various less developed countries around the world. The main problem is seldom a genuine shortage of funds. Rather it seems to be the absence of a real commitment to the task, lack of the organizational and technical banking skills needed to deliver credit to small farmers on such terms that they can make effective use of it, and the difficulty of loosening the tie between traditional creditors and borrowers at the village level.

We have not seen detailed studies of the need for or nature of rural credit in the riparian countries, or of the best measures to provide it. A survey conducted a few years ago in Thailand indicates that lending institutions are responsible for only 10 percent of agricultural credit. Of this, cooperatives supply one-tenth, so that for all practical purposes all but the largest farmers must be assumed to depend for credit on relatives and friends, or on village merchants and dealers—in short, commercial institutions other than banks.

Borrowing from friends and relatives is part of the "mutual aid" network described in Chapter 4 above and is without significant consequence for farm output. Most of it is extremely short term, and much involves transactions in kind or services.

The village merchant's lending—estimated at about a third of all agricultural credit—on the other hand, is closely tied to production, but only in the sense that the farmer tends to be in debt to the merchant to the limit of his marketable output. Most of the credit made available is for the purchase of consumption items rather than production inputs. In addition, however, the merchant provides a large number of personal services, for all of which he is reimbursed, in a sort of package deal, in the repayment of the debt. The repayment thus includes a large slice of money that is not an interest payment, and to that extent computed interest rates can be very misleading.

A more serious consequence of the close tie established by merchant lending of the "package" sort described above is that it leaves the indebted farmer wholly without credit-worthiness for any other loan transaction. Moreover, the farmer's first repayment priority would always be to the merchant, given his importance in matters other than lending. It follows that an increase in loan funds would not by itself increase lending. By the same token, a degree of credit-worthiness would be established if the noncredit functions of the merchant's lending could be transferred to some other activity or entity (such as more mutual help, village community organizations, savings institutions, spreading of information by government so as to attain greater diffusion of knowledge, etc.) and if some of the farmer's marketing could be carried on independently. In that event, the farmer's marketable surplus could serve as the basis for production-oriented credit instead of being a security for services rendered by the merchant.

There should be no illusions about the problems of installing new rural credit institutions widely throughout the region. The Bank of Agriculture and Agricultural Cooperatives, established in Thailand in 1966, engages mostly in short-term operations (up to one year) and has about three percent of the farm families of Thailand's poorest region—the Northeast—as clients. The bank charges 12 percent interest per year. As this illustrates, the cost of extending credit to many small farmers is likely to be high, and the whole enterprise is likely to be subject to considerable risk. Shortages of skilled banking personnel and physical facilities are possible. Negotiating and administering many loans, even when the average amount involved is small, may be very difficult, given the unfamiliarity of both borrowers and lenders with proper credit procedures. Moreover, shifting from well tried, even though primitive, production techniques to modern, more complicated ones inevitably entails increased risk of crop failures, at least until the new techniques become familiar. Hence losses on loans to innovating farmers may be relatively high despite the farmers' best intentions to repay.

Despite the difficulties, the effort to build greatly strengthened rural credit institutions must be made. No country has ever made the transition from primitive to modern agricultural techniques without such institutions. The Basin countries are unlikely to be exceptions to this rule.

Knowledge. Farmers in the Basin have developed, over the generations, many ingenious and effective ways of producing agricultural commodities under the particular climatic and other conditions in which they live. But modern agriculture has gained new dimensions through the application of science and technology; new understandings of the basic factors in agricultural production must be translated into specific operational methods to fit the circumstances of the various parts of the Mekong Basin. It is easy to talk, in very general terms, of new crop varieties, or new methods of water management, or greater use of fertilizers; but the farmer needs to know whether these elements are in fact appropriate for his location and he needs to know precisely how to apply these various general concepts to his particular situation.

The need for better information is nowhere greater than in the field of water management on agricultural lands.[6] There has been much written about irrigation in the Mekong Basin; but water management involves far more than irrigation in an arid climate; under monsoon conditions, there is likely to be far too much water for optimum crop growth at some seasons, as well as far too little at other seasons. Moreover, there will be years when there is far too little water at specific dates during the season in which there is normally far too much. A sophisticated water management system must be capable of preventing flooding from other areas, of disposing of excess water which falls on the land, of draining undesired water through the soil, and of applying additional water—all as needed, in whatever the amounts may be, and quickly. Still further, this is necessary not only on a macro or district scale but right down to the farm and to the field. The new rice varieties, for instance, produce their maximum output only when water levels on the fields can be controlled quite precisely (to a matter of a very few centimeters) at each date or stage in plant growth. An integral part of sophisticated water management is precise leveling of individual fields, so that the desired water depth prevails throughout.

The need for better information on water management illustrates the kinds of information that farmers require if they are to make effective use of modern production techniques. The question naturally occurs: what kinds of institutions are needed to provide farmers with the information they require? We suggest that the demonstration farm would be a particularly useful institution to serve this end. Farmers are pragmatists. They know instinctively, and frequently from hard experience, that experiment station results obtained under "hothouse" conditions are difficult to obtain under the more diverse conditions prevailing in their own

6. See Ch. 3 for an exposition of the problem.

fields. Hence it is important to demonstrate that new production techniques will pay off under those same conditions. Pilot projects, ranging from single demonstration farms to projects embracing several hundred farm families, widely dispersed throughout the region, are an effective means of doing this. Efforts in that direction should be pursued as a matter of high priority.

Such pilot projects must be backed up by centers of advanced science, including laboratories and research plots. Instances will arise in which the desired results will not materialize for reasons that are not immediately obvious, and in which specialists on some particular aspect of production must be consulted. But there seems little reason for these research centers to undertake genuinely innovative or original research; there are many research stations around the world, more ably staffed and financed, where the pioneering research can be done. Adaptation of such research to conditions in the riparian countries of the Mekong is an important and difficult enough task to absorb the full competence and resources of all agricultural research efforts within these countries.

Ideally, then, there would be scattered about the countries a number of pilot projects, which would develop close ties with both the more advanced farmers and the extension services; and a few laboratories and research stations, staffed with better trained specialists, whose main functions would be to cast the results of advanced research in other parts of the world into practical form for these countries and to serve as a technical back-up to the pilot projects. The whole would be integrated into a single system, with interrelationship among the parts suggested by the foregoing discussion. The system would include a single regional research center, ably led, with high continuity of financing and of staffing and with truly able scientists in all relevant specialities; under its control would be as many demonstration and testing farms or stations as necessary, considering both technical aspects and the need to reach farmers; the whole should be fully coordinated; and the riparian countries should resist temptation to establish for themselves or to allow outside groups to establish any agricultural research program which was not part of this coordinated structure. The demonstration farms would have direct ties to agricultural extension, whose personnel should in many cases be required to work directly on the demonstration farms to acquire practical farming competence, which is now nearly universally lacking among such men; or the pilot project might have its own staff of specialists.

This ideal arrangement may have to be modified in practice. Perhaps the countries of the region would be unwilling to have a single agricultural research system serving them all; perhaps each would insist upon its own. This would be unfortunate, particularly in the cases of Laos and Cambodia. Both countries lack, and will lack for many years, a competent and adequate corps of agricultural specialists of their own; and their ability to attract competent foreign specialists, regardless of the agency for which they work, is low. Thailand has progressed further in agricultural research than the other countries, although the general criticisms made in this report apply to it, and perhaps it could go forward alone better than the other countries. But there would seem to be some gain for it in being part of a regional organization. More serious than these nationalistic considerations is the problem of bureaucrats entrenched within the research establishments, who may have had advanced training in other countries but who lack real professional and practical competence, and who are satisfied to continue in past patterns of activity.

The fact that emphasis has been put on institutions for imparting knowledge of modern production techniques and generally upgrading the management skills of Basin farmers does not mean that nothing should be done to raise their general educational level. As was noted in an earlier chapter, the situation with respect to general education in rural areas is very unsatisfactory throughout the Basin. Experience in the developed countries suggests strongly that eventually this situation will have to be improved if the Basin countries are to achieve sustained agricultural development. However, it does not appear that improvements in general education are a prior condition for setting the process of agricultural development in motion. There is an impressive body of evidence that even semiliterate, formally untrained farmers can make considerable progress on their own when they have the incentive to do so. The rapid spread of the new rice varieties in many parts of Asia is a case in point. The speedy response of Thai farmers to expanding markets for corn is another, although this apparently did not involve very considerable changes in technique.

The stress on improving management techniques, rather than general education, therefore, results from a judgment of development priorities. The evidence indicates that even with the present low level of general education Basin farmers can learn to employ more modern techniques if they are shown how to do so. Moreover, the payoff to these techniques can be very large and relatively fast in coming. It makes sense, therefore, to capture these gains first. Then, as the economy of the region grows, increasingly greater weight can be put on improving the level of general education in rural areas.

A Note on Improved Health Services. The public health services provided to Basin farmers are very poor, and a variety of seriously debilitating and chronic diseases are prevalent in the area. We are not in a position to evaluate this situation or recommend specific measures to correct it. It is worth noting, however, that aside from their contribution to the general well-being, investments in expanded and improved public health services may yield a substantial economic return. It seems likely that the ability of farmers to work is now significantly impaired by disease. A general improvement in health might thus result in a significant increase in production. What, for example, would be the effect on production over the next decade of public health measures permitting a three percent annual increase in the physical energy expended by the farm labor force, assuming economic incentives to put out the additional effort? To ask the question is to suggest that the economic payoff to public health investments may be very large indeed. A systematic investigation of the issue appears highly worthwhile.

CHANGES IN COMMERCIAL AND VILLAGE INSTITUTIONS

Chapters 3 and 4 indicated that weaknesses in the institutions providing input and marketing services to farmers inhibit the adoption of modern production techniques. While government agencies perhaps have a role to play in strengthening these institutions, there is reason to believe that private initiative may be more effective. The essence of efficient input and marketing services is fast and flexible response to sometimes rapidly changing market conditions. Government bodies generally, for reasons noted in chapter 4 above, and those in the riparian countries particularly, are not noted for these characteristics. On the other hand, Basin country "middlemen," on both the input and product sides of the market, have evidenced a keen eye and quick response to emerging market opportunities: witness the experience of Thailand in responding to the rapidly growing market for feed grains in Japan.

One of the obstacles to reliance on private initiative for input and marketing services—and one that will not be overcome by pretending that it does not exist—is that these services are predominantly in the hands of Chinese merchants. There is little detailed information on their role in agriculture or their day-to-day relationships with farmers, although it is known that rice millers, for example, as explained earlier in this chapter, provide credit and other services to farmer-clients. Affording greater social recognition or social acceptance to smaller Chinese businessmen could greatly assist them, and the farmers as well, to arrange transactions on a much larger scale. As they become more accepted in the larger society, they might rely less completely on mutual relations among themselves, thus becoming still more acceptable in the larger society. If smaller Chinese merchants become as integrated as the largest ones already are, they can participate more fully in providing commercial services to agricultural modernization. The suspicion which both farmers and national officials have felt toward the alien Chinese merchant has prompted most farmers to treat him with distrust and most officials to circumscribe his activities with elaborate controls. But a more effective way, for example, of enforcing fair agricultural prices than through controls, would be to open up the commercial distribution of agricultural inputs to all interested firms, perhaps giving them initial assistance but placing them in competition with international supply firms and with a few government agricultural services. Moreover, anything that would divorce marketing of farm products both from marketing of inputs and from supply of credit would help. The capability of the resident alien businessmen not only in the riparian countries but in Southeast Asia generally to make a profit in very constricted circumstances has impressed many outside observers. The opportunity to participate more extensively in agricultural commerce, perhaps reinforced with import licenses, may be all the reward necessary to cause them to assume a much more active economic role.

There is a real question whether the issue can be further probed *in vitro*. Is it possible to design a pilot project in a specific location in which new commercial channels of moving inputs and outputs can be tried out? Because of their importance we would strongly favor an attempt in that direction, borrowing perhaps from techniques that have recently been developed to test rather than to guess at the reactions of people to specific new institutions (such as, for example, guaranteed minimum income experiments in the United States). Such projects might also help to overcome the widespread reluctance to grasp this particularly spiny nettle.

Productive, regular use of an elaborate set of agricultural inputs by farmers would require of them new knowledge and continuous access to sources of information in order to test and evaluate their results. A stumbling block in effective delivery of agricultural services to village farmers is the contrast between the formal, impersonal, official associations established by government officials for village farmers, on the one hand, and, on the other, the informal, personal groupings in which village farmers help each other in rice cultivation. Despite many optimistic reports written by specialists on cooperatives, the record of farmer associations in the Mekong Basin is very poor. Traditionally, farmers have banded together in ad hoc groups at the particular times and places that served their mutual interests. Thus far many of them have not seen how a government sponsored farmer association, with its formal organization and very limited resources, would be of any substantial use. In the words of the *Asian Agricultural Survey*, "the successful establishment of farmer organizations can only occur after farmers have gained enough personal experience with new technologies to recognize the gains that can be realized through group action." Experience in the riparian countries is a case in point. If most of the supplying of agricultural inputs were handled as a commercial venture, the private firms involved could deal with individual customers as easily as with groups of farmers. Thus, very little organization of farmers would be necessary for the new rice cultivation, except in such groups as arise spontaneously. On the other hand, farmer associations for marketing products might, and groups of farmers organized to maintain irrigation projects would, become essential.

GOVERNMENT PRICE AND IMPORT POLICIES AFFECTING AGRICULTURE

Farmers will not adopt new, improved inputs unless it is to their economic advantage to do so. This means that input/output price relationships must be favorable and that the inputs must be available in the amounts needed. Hence government price policies affecting agriculture, and trade policies, particularly regarding imports of fertilizers, pesticides, farm machinery and parts, fuel, and so on, are of major importance. To the extent that the region's products move in world trade, and domestic markets are exposed to imports, Basin farmers are exposed and must adapt to world market conditions. Government policies can significantly alter the pattern of prices as perceived by farmers, however. The Thai government's policy of extracting a "premium" from the price of rice is an example of this. The price received by Thai farmers is significantly less than it would be if their production were sold at the world market price. While this may have been justified as a form of taxation when prices were high, in view of the probable decline in rice prices the Thai government may wish to reconsider its policy, particularly with respect to its effect on the incentives of Thai farmers to adopt the new rice varieties and generally to modernize their rice production technology.

Maintenance of favorable input/output price relations may give farmers the incentive to adopt more modern practices. They may still fail to do so, however, if the necessary inputs are not available in the requisite amounts. Since many of these inputs—fertilizers, tractors, and so on—are wholly or in large part imported, government import policies can have a major impact on the pace of farm modernization. The problem is a common one throughout much of the underdeveloped world. Most of these countries face more or less severe balance of payments problems arising from overvalued currencies and strong demands for imports of both capital and consumer goods resulting from the pressures generated by development itself and the rising expectations associated with it. The response commonly is the imposition of import controls. The priorities of government officials and their ability to enforce them will determine the distribution of imports between capital and consumer goods. The danger that the share of capital goods will be less than that needed to achieve sustained development is real. Agricultural inputs must compete with other capital goods within a total which may already be inadequate. Since agriculture, at least until recently, has had less "prestige" among planners and policy makers than industry and other non-agricultural activities, the position of agricultural inputs in the competition for imports of capital goods has frequently been relatively weak. Consequently farmers have not always been able to buy these inputs in the amounts indicated by the pattern of input/output prices.

The governments of the four riparian countries will of course set their own import policies in response to a variety of conflicting interests and pressures. There is no intention here to indicate the priority which should be given to imports of agricultural inputs. It is important for the various governments to be aware, however, that measures to strengthen the institutional and physical infrastructure serving agriculture and to establish favorable input/output price relations for farmers will all fail if import policies limit the supply of inputs below the amounts needed to implant and sustain modern agricultural practices.

THE IMPORTANCE OF ADMINISTRATIVE REFORM

Previous sections of this report have indicated the major changes necessary to transform Basin agriculture: through appropriate credit, price, and import policies make available greatly increased and reliable supplies of economically priced agricultural inputs; improve water management systems to obtain a much higher level of water control; establish experimental-demonstration-extension programs to provide momentum and guidance for modernizing agriculture; and improve institutions of commerce and village life. These changes will in turn require fundamental changes in the functions performed by, and substantial increases in the capabilities of, the institutions of government. These changes cannot be achieved without considerable alterations in the system of rewards in all these institutions. Because these are highly complex, we indicate only briefly the types of changes which should be further explored with the governments and people concerned.

Not as yet fully competent to carry out the demanding functions listed above, all the governments will require extensive foreign assistance; but unless the governments impose upon themselves profound administrative reforms, that assistance can lead only to further structures that are monuments without wide economic impact. Among such reforms would be measures cited below that aim at (a) removing the administration of agriculture from a centralized hierarchy; (b) reducing the motivation for the practice of corruption; and (c) modifying the current elitist rule.

In Chapter 4 we pointed to the persistence of relations along lines of centralization and hierarchy. To overcome these, new rewards must be established in government that will—

- result in delegation of authority more closely to the units of activity;
- increase the numbers of centers of decision making and innovation;
- create a greater stress on efficiency of performance than on its appearance;

provide more adequate feedback guidance to central administrators;

narrow the very wide social gap between urban officials and village farmers; and

associate job promotions more closely with caliber of performance than with personal relations with superior officials.

To minimize corruption, new rewards must be sought—not necessarily only of a monetary kind—that would enable officials both to make an adequate living from their jobs and to take greater pride in them, and new impartial measures for punishing corruption wherever it appears must also be sought. In any event, what is important is to identify the motives in order to be able to fashion or favor valid substitutes. It is easy to be stern in one's view of corruption, but sternness must be tempered with an understanding of the origins of corruption. An interesting comment made recently by H. E. Soedjatmoko, Ambassador of Indonesia to the United States, in the paper cited in Chapter 4 (see n. 15), is worth quoting in that connection. In cultures, he says,

> where the family is the most important social unit in society, and where family loyalty and solidarity are virtues of the highest order, the growth of an effective feeling of superseding and overarching loyalty to the nation, essential to the solidification and effectiveness of the new nation state, may be difficult to bring to life, even where its need may be intellectually conceded. The persistence of "corruption" in some nations often reflects the incompleteness of this transition to the more impersonal organizational requirements of the modern nation state.

Fundamental to these issues of hierarchy and corruption is that of elitist rule. The elite national leaders, who enjoy the highest status and income from the current social and economic order, are the only ones with enough authority to initiate the reforms needed for modernized agriculture, but these reforms would reduce the favorable positions which they and their subordinates currently monopolize.[7]

Thus a basic paradox in agricultural modernization is that top government leaders, who are the major spokesmen for modernization and development of their countries, would have the most to give up in a thorough program of agricultural development which reached most of the farmers—a situation that, *mutatis mutandis,* prevails in any society faced with an urgent need for substantial transformation. A major reformulation of social and economic rewards could change this. It would contain, as a minimum, the following elements:

1. Increasing government salaries sufficiently that they will serve as sole source of income.
2. Enlarging the opportunities in business for members of national majority ethnic groups so that government officials will not feel compelled to enter alliances with business (see p. 35) in order to control ethnic minority entrepreneurs, and so that

7. Several observations pertinent to the rule of Asian elites appear in the *Review of the Social Situation in the ECAFE Region for 1970.* Asian traditional elites are being replaced by rising upper middle classes of bureaucrats (civil and military), entrepreneurs, and professionals. Although systematic, precise data are lacking, disparities of income are very large and seem to be increasing in most Asian countries. Incomes are normally the highest around the capital cities, where government, commerce, and industry provide the greatest opportunities for wealth. A survey in 1963 found average family incomes in the Bangkok area to be Baht 39,000, while cash income for 70 percent of the families in northeastern Thailand was only Baht 3,000. Even when the substantial noncash income of the northeastern farm families is taken into account, the disparity remains very great. The ECAFE study poses two basic social issues for national development: (1) Can the new elite provide technical and organizational leadership for development? (2) Will the new elite have the "commitment and will to lead broad movements of democratization which would draw the inert masses actively into the development process?" The ECAFE study finds the first issue already answered affirmatively but the second issue still remaining moot, since the new elites have profited most from the postindependent concentration of wealth and power in the capitals, and since the new elites in some countries have close interests with the old elites. This last relationship is a particular inhibition against extensive land reform.

members of the ethnic majority groups will have more adequate legitimate opportunities for wealth.

3. Making performance of specified developmental tasks of government more rewarding with salaries, promotions based on job performance regardless of location, normal fringe benefits, and recognition of special honor so that capable, ambitious officials can stop trying to avoid such jobs.
4. Establishing critical professional evaluation of performance and rewarding the provision of necessary information, even if negative, by field personnel to office headquarters.
5. Making it more rewarding to take initiative and to make decisions locally rather than wait for guidance from distant central offices.

The changeover to and emphasis on these rewards is a delicate, domestic affair, and the role to be played by foreign development agencies is neither obvious nor easy; nor can results be expected to show up rapidly. Moreover, the changeover is diplomatically risky; but so eventually is the conventional approach of training more skilled personnel or providing more equipment or structures in an institutional environment unprepared to use them. The task is delicate but quite possible with some steps like the following on the part of international development agencies, the Mekong Committee, and the riparian governments:

1. Examine the current performance of tributary projects in all the riparian countries where possible, to clarify the effects of economic conditions, social institutions, and cultural values on the administrative performance and physical results of these projects.
2. Collect ideas from all types of participants and observers in a project on ways to improve organization of the project.
3. Sponsor discussion of these ideas with a wide range of officials in agencies concerned.
4. From these discussions, formulate, with the agencies concerned, programs of further assistance involving pertinent administrative and social reforms.

5. Include such reforms as legitimate projects for development assistance.
6. Offer further assistance to Mekong development projects on condition that these reforms be carried through.

A paradox of modernization was stated above: the top leaders, who would be instrumental in making the changes, have the most to lose by them. The opportunity of modernization, however, is that they too can share in the generally higher level of wealth. Given the structure of their societies, they can initiate many of the reforms needed. International assistance can make such reforms more attainable by making them less expensive, and more legitimate and by enabling the government to alter the old system of rewards to the participants. Such a new approach on the part of international agencies would require the mobilizing of some of the best riparian and international social scientists to examine current and projected Mekong projects. It would also require further training of riparian social scientists. As riparian professionals take greater part in the assessment of the social modernization aspects of Mekong development, the entire task will shift from being delicate toward being difficult, but possible.

This discussion suggests that the success of the agricultural development strategy proposed here may ultimately depend upon the success of the riparian countries in instituting administrative reforms. The strategy assumes governmental initiative in removing key institutional and physical infrastructure bottlenecks now inhibiting the transformation of agriculture. It is doubtful that these initiatives will succeed or in many cases even be taken, given the present modus operandi of most governent agencies. Fundamental reforms are essential.

The *Asian Agricultural Survey* made this point well, though one need not agree with every detail or the exact wording:

> Efforts to improve administrative procedures as they affect agricultural development without engendering basic changes in the whole structure of governmental operations will not have much impact. Agricultural activity is just one among many within the interlocking work of the government, and no

sweeping reform can be made handling this element without corresponding reforms in the methods handling the others. Yet, unless administrative reforms are made, there is little evidence that future national programs for agriculture in most countries will be conceived, implemented and operated with any greater success than past programs.

ACHIEVEMENT OF PRODUCTION GROWTH TARGETS

The foregoing discussion implies that a development strategy designed to remove bottlenecks in the institutional and physical infrastructure serving agriculture will permit the Basin to transform its agriculture and achieve satisfactory growth of output. With respect to physical works, the proposed strategy emphasizes investments in numerous relatively small-scale projects designed to improve the operation of existing systems rather than in major new projects, such as those proposed for mainstem development.

In an earlier chapter it was asserted that agricultural production will have to grow at about 4 percent annually over the next couple of decades if total income growth targets for the region are to be achieved. It remains to ask whether the suggested investment strategy is consistent with 4 percent annual growth in agricultural production. The period considered is the next two decades, or to 1990. This span of time is selected because we think it long enough to permit the region fully to develop potential tributary projects and to make the infrastructure and institutional improvements needed to set in motion and sustain the process of agricultural modernization. Moreover, as is shown in Chapter 6, the optimum time for bringing mainstem projects into operation is no earlier than 1990.

According to the ABP Report, annual average crop production in the region in 1963–68 was 10,776,000 metric tons of paddy and paddy equivalent of all major crops (combined on the basis of calorie content). We assume that crop production must grow at 4 percent annually from 1963–68 to 1990 to achieve growth targets. Hence crop production in 1990 must be 29,310,000 metric tons.

In the ABP Report, the planted area in 1990 is projected at 10,903,000 hectares, and 1,100,000 of this could be irrigated (double-cropped) with tributary projects alone, i.e., no mainstem development. We assume that this double-cropped area will yield 6.0 tons of paddy or paddy equivalent in 1990. This appears to be a quite conservative estimate. The yield is less than yields already being obtained on fairly extensive double-cropped areas in the region.[8] If the Basin countries adopt the various measures recommended elsewhere in this report, yields of 6 metric tons per hectare per year on irrigated, double-cropped land should be easily attainable by 1990.

Total production on irrigated, double-cropped land, therefore, is projected at 6,600,000 metric tons of paddy and paddy equivalent (6 tons per hectare times 1.1 million hectares brought under irrigation by tributary projects). This implies that production on nonirrigated land must be 22,710,000 tons in 1990 (29,310,000 tons total production less 6,600,000 tons produced on irrigated land). According to the ABP Report projections, nonirrigated land will come to 9,803,000 hectares (10,903,000 hectares minus 1.1 million irrigated hectares). The yield of nonirrigated land, therefore, must rise to 2.32 tons per hectare in 1990 from 1.31 tons in

8. Vernon Ruttan writes as follows: "Both in the Philippines and in Thailand most of the lowland rice is grown during the rainy season without irrigation. Under this rain-fed system of cultivation village or provincial average yields rarely exceed 1.5 metric tons per hectare. However, in fully irrigated areas in both countries, such as Chiengmai in Thailand and Laguna in the Philippines, it is not uncommon for average yields to exceed 3.0 metric tons in the wet season and 3.5 metric tons in the dry season over fairly extensive areas. On individual farms, such as those participating in contests or those under experimental conditions, the yields of the same varieties frequently fall in ranges of 4.0 to 4.5 metric tons in the wet season and 5.0 to 6.0 metric tons in the dry season. . . . New varieties now being developed appear to have yield potentials within ranges of at least 6.0 metric tons during the wet season and 8.0 metric tons during the dry season." "Strategy for Increasing Rice Production in Southeast Asia," in W. W. McPherson, ed., *Economic Development of Tropical Agriculture: Theory, Policy, Strategy and Organization* (Gainesville, Fla., 1968).

1963–68, an average annual rate of increase of 2.4 percent.[9] This is a rapid but by no means unattainable rate of yield increase. Between 1948–52 and 1963–67 rice yields in the Basin rose at an average annual rate of 1.7 percent. This occurred despite the war-induced disruptions in this period and in the absence of widespread innovations in production techniques. As noted earlier, except in South Vietnam the new rice varieties had little impact on the region by the late 1960's, fertilizer consumption was quite low, and water management systems were generally inefficient. If the Basin countries move toward adoption of modern agricultural practices, acceleration of the rate of yield increase on nonirrigated farms from the current rate of 1.7 percent annually to 2.4 percent should be quite feasible.

Therefore, achievement of 4 percent annual growth in crop production over the period to 1990 without construction of mainstem irrigation projects seems well within reach. Even if the rate should fall somewhat short of this, however, *total* agricultural production, i.e., crops plus animal products, still could grow at 4 percent annually. Output of animal products, while small at present in relation to total production, is expected to grow much faster than crop output if the region adopts the various measures suggested elsewhere in this report. Hence total agricultural production could grow at 4 percent annually over the next couple of decades even if the growth of crop production were somewhat less than this.

We conclude that a development strategy focused on tributary development and improvement in the institutional and physical infrastructure serving agriculture will permit farm output to rise at a rate consistent with achievement of total income growth targets for the region.

9. Little is gained from slicing such conjectures very fine. For example, a 2.5 percent rate would be arithmetically compatible with a yield of only 5 tons per hectare on irrigated land, a level which some consider more realistic than 6 tons. The point is that various reasonable combinations can add up to the stipulated total.

6

Mainstem and Tributary Projects: A Question of Sequence

In the discussion in Chapter 5 of an investment strategy for the region it was argued that for the next two decades principal emphasis in water resource development should be placed on improvements in existing water distribution systems and extension of the irrigated area through tributary development rather than on mainstem projects. The question of when to begin mainstem projects is of major significance and hence deserves much further treatment than we have given it so far.

The issue fundamentally concerns the proper sequence of investments as between mainstem and tributary projects. Efforts to resolve this type of issue conventionally focus on such criteria as benefit/cost ratios, engineering efficiency, and the like. For the Mekong a wider range of considerations seems appropriate. We believe that cultivation practices as well as the system through which commodities reach the consumer must be improved before the riparian countries can enjoy the full agricultural benefits from mainstem projects. We believe that available budget and development expertise should be allocated to a number of smaller projects which will provide opportunity for improving the practices and supporting institutions of Mekong Basin agriculture. In addition, we find the tributary projects likely to lead to wider and earlier diffusion of benefits from the funds to be invested.

To begin with some major considerations, it seems to us that the amount of capital required by mainstem projects argues against their early initiation. The cost of Pa Mong alone is estimated at $1.1 billion, which is equal to the average annual total investment in Thailand in recent years or to one-fourth of its average national income. Such massive investments for single projects greatly affect the capacity of the country to make other investments—either from internal savings or from borrowed capital, possibly for many years to come. As a consequence, flexibility of investment and of economic policy is materially reduced, as is the capacity to respond to unforeseen future economic conditions. Such massive investments for a single project may also affect the flexibility of international lending institutions: they may find that substantial supplemental investments are required to secure the originally anticipated benefits.

This is not to suggest that the size of investment required for tributary projects is small: investment for the seventies is set in the ABP Report at over $1 billion. This is a large amount; but it represents capital applied to a number of smaller projects, for each of which there is a range around the optimal level of investment within which the project is economically viable. Thus, while the amount of capital required for tributary projects may equal the cost of Pa Mong, the former is a more flexible investment package.

Tributary projects also appear to offer more opportunities than do mainstem projects for experimentation with various types of development investment. Promoting economic and social development is an inexact art, and some opportunity for trial and error should be built into the early phases of a development program so that the design of later phases can benefit from the lessons learned. This could prove particularly useful in making discernible the kinds of institutional innovations needed to assure full utilization of water projects.

A third general point is that there is a far greater opportunity for upgrading the skills of the local population in the course of construction of the smaller tributary projects than in a project of the caliber of Pa Mong. Indeed, experience on Mekong tributaries indicates that the Thai have been able to use project construction and operation to increase such skills, of both engineers and artisans, while the Lao have been obliged to increase their dependence on foreign skills. We take the view that economic development entails raising the capacity of the population, and that this involves the acquisition of new technical competence and skills. The opportunities for this afforded by the construction of several local projects as opposed to one large one (which in all likelihood would require a sizable complement of foreign manpower) appear to be an important additional advantage of a tributary-project focus for Mekong Basin development over the next couple of decades.

Even if one could secure general agreement to the above principles, it would still be quite possible that economic or political needs would alter preferences based on these principles. For example, a market might develop for commodities which the Basin countries could produce only if substantial amounts of new irrigation water were available from mainstem projects. Or possibly tributary projects could not be effectively undertaken for engineering reasons, unless one or two mainstem projects were also built, or agriculture in the Delta could not be modernized without the construction of Pa Mong or other mainstem projects. It has been alleged that hydroelectric power from Pa Mong is essential to supply the needs of Bangkok as well as the generally growing demand for electric energy elsewhere in the riparian countries. And finally, there is always the possibility that political reasons might make it advisable if not essential to start construction of mainstem projects. We shall take a look at each of these issues.

We concluded in Chapter 2 that domestic and foreign demand for the region's agricultural production could be expected to grow at about 4 percent annually. In Chapter 5 we argued that this increase in demand could be satisfied by an investment strategy focused on improving the institutional and physical infrastructure serving agriculture and on tributary development to expand the irrigated area. Of course, if demand increased much faster than 4 percent annually, this strategy might not suffice. Given the apparently bearish long run prospects for the world rice trade, however, we think more rapid growth in demand unlikely. Thus we cannot see any need for irrigation from the construction of mainstem projects for many years into the future.

It is on the other hand imperative that modern water management, appropriate to local conditions, be introduced and mastered by agricultural producers in the Basin. This will take time and effort; but it will redound to the advantage of agricultural production during the balance of the century and will be essential to the success of mainstem projects, when they are constructed in the future. This is especially true if one considers diversification of output for both domestic and foreign markets. It will take time to establish production practices as well as outlets for the output. On the other hand, there is need for projects with a short payoff horizon, if only to sustain the momentum of this very ambitious and protracted undertaking. Tributary development can fill that need.

Much emphasis has been placed on the future need for electric power, especially in the Bangkok area; and indeed reported load growth, above all in Thailand but also in South Vietnam, has been exceedingly rapid. Load growth projections of 20 percent per year and better are substantially higher than they are almost anywhere else. It has not been possible to analyze the validity of the projected rates for the Basin, and it would thus be inappropriate to challenge these data. If the projections are accepted, then the projected load must be met one way or another; and the justification of a hydroelectric project will rest on a comparison of the cost of power from Pa Mong with the cost of power from alternative sources—in this instance, thermal generation.

From data prepared by Acres International, a Canadian consulting firm with experience in the area, and made available by the World Bank, we can draw some useful comparisons. These point sharply to the conclusion that (1) power generation should not by itself set the direction and sequence of Basin development; and (2) the benefits of Pa Mong from power generation would be greatest if postponed to the last decade of the century, unless the cited load projections underestimate growth—not a likely contingency. This conclusion holds throughout a range of reasonable future fuel costs: 20 to 40 cents per million Btu. That is, to be competitive with alternative thermal power in this fuel cost range, the price of Pa Mong power would consistently permit a higher internal rate of return if built later rather than earlier. Specifically, if the date of Pa Mong availability were postponed from 1983 to 1993, the competitive price would allow the rate to rise from 8 to not quite 10 percent for 40-cent fuel, from 7½ to 8½ percent for 30-cent fuel, and from 6 to 7 percent for 20-cent fuel.

The foregoing analysis considers neither any future reduction in fossil fuel costs nor any likely future technological change in thermal power generation, apart from fuel prices. However, we judge that as little as a 10 percent cost overrun on Pa Mong would spell a reduction of 0.75 percent, or more or less a proportionate decline, in its internal rate of return. Finally, one must point out that no rate of return lower than 10 percent appears realistic considering returns available from alternative uses of capital. Such a rate is just barely approached in the variant of construction in 1993 at the high fuel cost of 40 cents per million Btu.

As one considers the power issue further, one moves to considerations which cannot be asserted with full certainty. For example, ecological consequences of mainstem projects are likely to be far reaching and their putative costs must be stacked up against the estimated benefits. All one can say is that they probably would further tip the scales in favor of power and irrigation development elsewhere. Creation of lakes as vast as would come into existence behind Pa Mong can no longer be viewed with relative indifference. But in the time at our disposal we are able only to put in a caveat here and point out at least the direction in which appraisal of mainstem development would be affected.

As stated above, we have accepted the power load projections, regarding both rate of growth and customer composition. It is difficult to believe that the rate of growth is understated. Any error is more likely on the side of exaggeration. Again, a lower rate of growth would diminish the need for building Pa Mong at an early date.

The uses to which additional power is likely to be put are worthy of some comment. Data for Thailand suggest that the bulk of the new power supply would go not to residential but to industrial use, just as industry is now the major user, accounting for two-thirds of total consumption in the Bangkok area and little less than one-third elsewhere. Thus one cannot easily argue that the additional power would go primarily towards providing added comfort and even luxury for the well-to-do in the cities, especially Bangkok, while neglecting the well-being of the bulk of the people living in the rural areas. On the other hand, the additional power would fail to bring significant improvement in the drastically uneven distribution of electricity between the regional aggregates for which these figures are available. By 1978, Thailand's central region, which means predominantly Bangkok, would consume about 1,100 kwh. per person (much more, of course, per customer), while the rest of the country, containing 70 percent of the country's population, would consume only 110 kwh. per person, or one-tenth as much. Thus, the contribution to well-being from greater power consumption would have to come largely via greater benefits from industrialization elsewhere in the country if a more equitable diffusion of benefits were to be demonstrated. We have not attempted either to prove or to disprove such a thesis, but we wish to call attention to the need for evaluation of benefits from increased power generation.[1]

Even though load projections show the continuing importance of industrial consumption, residential use too would grow steeply, from 1,700 kwh. per customer in 1969 to 3,700 kwh. in 1978, in the area served by Thailand's Metropolitan Electricity Authority. This consumption would be enjoyed by fewer than half a million customers, accounting perhaps for 2.5 to 3 million people out of the country's projected 46 million inhabitants towards the end of the seventies. Together with the closely associated "commercial" category this use would absorb about one-third of all power distributed by the Metropolitan Electricity Authority in 1978,

1. Contributing to our concern is a perusal of the *Report of the Electric Power Management Advisory Team on Vietnam: Final Report, 1967,* which in its comments on load growth makes much of additional air-conditioners, clothes washers, and other domestic appliances.

accounting for nearly one-quarter of the country's consumption.[2]

Altogether, there are great differences in impact upon development between, and even within, broad classes of use to which power from mainstem projects could be put. This is not, of course, to disparage either the usefulness of specific power-intensive industries, or power projects as such, if water controls of the kind that would make the power project merely an adjunct, or would permit it to become so at a later date, are required in any event. All we are saying is that the *volume* of power demand represents only one dimension of its significance for development.

One must inquire next whether the financing of Pa Mong is an attractive early development investment in terms of its potential for providing a water supply for agricultural services. Here one must fall back on the discussion regarding the chances of obtaining increased agricultural output from alternative sources (Chapter 5). If new agricultural practices will not produce results in the service area of tributary projects, there is no reason to assume they will do so in the service area of mainstem projects. Conversely, if they yield results elsewhere, the contribution from Pa Mong, or other mainstem projects, will not soon be needed. To

2. The issue has additional dimensions which we can only touch on but which need to be explored with some care. There is little doubt that increased power supplies would hasten the pace of urbanization in Bangkok and other large cities as opposed to that in rural towns. Modernized agriculture, especially when associated with small scale processing industries in the countryside, could make a sustantial contribution to mitigating the disparity between the rapid annual increases of population in the large cities and practical stagnation in the rural towns, whereas additional power for cities would tend to aggravate it. Other aspects of the issue that need investigation are the diffusion of benefits from types of industrial power use envisaged, and the extreme inequity imposed upon farms and villages flooded out by Basin reservoirs associated with power generation and the dismal results thus far from attempts to make them participant beneficiaries rather than passive victims of development. Special attention and effort need to be focused on this particularly difficult aspect of Basin development. In this context one cannot view with anything but grave concern the prospect of resettling an estimated half-million people that would be displaced by the proposed Pa Mong and Stung Treng projects. Much further work on current and future tributary projects seems a prerequisite to tackling that particular task.

put it sharply, the earlier one builds Pa Mong, the less attractive it will be as a power producer—as pointed out above—and the less time there will be to prepare the ground for making it an effective element in the Basin's agriculture. Thus postponement recommends itself from both points of view.

The fourth and final question we must ask is whether there might be political constellations that would make an early start of construction on the mainstem projects advisable or perhaps even necessary. At the moment certainly the shoe is on the other foot. Any construction now would be unduly risky. Many parts of the Basin were as of mid-1970 inaccessible even for surveying and planning teams; construction would pose nearly impossible problems in many areas, given the need for access through the riparian countries; and continued uninterrupted operation of projects is unlikely, were they to be built. Under these circumstances it is highly doubtful if any outside investment would or should provide the necessary capital—and only outside investors have capital in the amounts required.

Secondly, any construction of mainstem projects would require some kind of treaty or formal understanding concerning the division of costs and benefits from upstream development. Indeed, some top agency that encompasses at least the four riparian countries would be needed. The arrangement could include only Pa Mong and, at the minimum, could provide for agreement between Thailand and Laos on construction of the dam and for expression by the downstream riparian countries of no objection. The latter countries would, however, benefit considerably from operations of Pa Mong even if it were operated solely to maximize power output (an event which we do not favor; see above).

Economic efficiency, as well as equity considerations, suggests that the downstream riparians contribute to the cost of river regulation in some proportion to the benefits received. The same applies in principle to benefits from Stung Treng. Difficult as it may be to negotiate a treaty covering these unrelated points before constructing any mainstem project, construction of such a project prior to conclusion of a treaty would be even more difficult and more fraught with possibility for discord.

One cannot, however, exclude the possibility that disorder and hostilities in the area may soon begin to decline sufficiently to cease being obstacles to construction activities. Specifics apart, outside agencies must be prepared to move as political events indicate. We can see three basic variants. Continuation of hostilities at current levels will continue to restrict both on-the-spot investigation and construction, but need not hamper activities that are not location-bound, certainly not establishment of pilot projects, upgrading of research and extension, or steps toward better utilization of completed projects. A gradual reduction in unrest without a formal settlement would permit intensification of work on smaller projects. Finally, a formal settlement is likely to find the United States government—and perhaps other potential donor countries—taking increased initiative and willing to provide more capital to the point where Mekong River Basin development, mainstem projects and all, could become a major element in a peace settlement. This might be especially true if such a settlement embraced economic aid to North Vietnam, an event that might even alter the evaluation of hydroprojects; at least that possibility is worth exploring.

Moreover, regardless of the timing of mainstem construction, there are essential financial and organizational steps to be taken in the near future (see Chapter 5). These relate to demonstration-research-testing facilities, including pilot projects, crop diversification, market research, the full use of water projects now existing and under construction, and the pursuit of design studies for one or more mainstem projects. These will flow in specific time patterns, with pilot projects coming in at once; the full utilization of subbasin projects being also tackled now but with results spread over a longer time; and, as stressed above, mainstem research and engineering being a continuous activity with the longest time horizon in terms of actual facilities emerging.

If an international settlement were reached, it is thus not at all unlikely that to seal the agreement large international funds would become available, much larger than could be absorbed by tributary projects alone. Indeed, it is quite likely that political considerations arising out of a settlement would become overriding and face the planners with the need for action—the best action feasible at a time set not by them but by "events." Even if other factors were not the most favorable, it would be important that the opportunity be grasped intelligently. If for no other reason, studies on planning including mainstem projects should be pushed at a rapid rate. Specifically, detailed engineering studies and planning for Pa Mong should be pursued in parallel with the action programs that have much shorter payoff horizons.

There are other large gaps in information and planning. Among them is poor coordination of power projects with nuclear or fossil fuel power facilities; there is insufficient knowledge of the interrelationship with flood control, irrigation, and water management for the Delta; the role of some major projects, notably Tonle Sap and the Great Lake, needs to be elucidated; more emphasis needs to be given to the economics of timing of various projects in relation to one another and to well specified economic needs; and no explicit study has come to our attention of the engineering feasibility or economic advantages of diversion of water *outside* the Mekong Basin.

These are only examples of what still remains to be done in order to take advantage, if necessary, of early availability of large funds. But even should such a contingency not arise there is much to commend long-range planning; the fact that internal capital is scarce, that available external funds may be inadequate, that governmental and private entrepreneurial competence and capacity are limited, that social change will be difficult enough to stimulate at best—all these argue for the most careful selection of mainstem projects and their characteristics. A bad choice is likely to mean not only wasted resources but lost opportunity to do better elsewhere.

In summary then we would say that the preconditions for commencing mainstem projects are:

1. that substantial results—physical, managerial, and social—have been reached in the tributary projects;
2. that markets are clearly opening up in a sustained way for absorbing the increased output that would come from the development of mainstem projects; and

3. that associated power production is carefully channeled so as to minimize the chance of its accentuating social and economic inequities both between rural and urban areas and within urban areas.

Even if sudden availability of large funds may make it necessary to override to some degree some or all of these factors, they should never be lost sight of.

One difficulty of immense proportions is common to all projects, though more pronounced in the context of mainstem projects, and that is that the farmers of the Basin countries have little tradition or experience in working together cooperatively, except at the most local level. They are not accustomed to exercising political power; indeed, they possess little or none. Nor do they look to their national government for a solution to their practical problems. Each of a plethora of farmers' "cooperatives" or committees is sponsored and dominated by a government agency, which plans and carries out resource development and agricultural programs. Sometimes efforts are made to inform local people, sometimes not; but the local people are never genuinely involved. Under these conditions, there is grave danger that any major Mekong project will become identified in the minds of the riparian governments and of local people as an enterprise of the lenders or other outside groups, rather than as an enterprise of the riparian countries concerned. It may be viewed as "theirs," not as "ours." If the project encounters serious difficulties or if it fails, the responsibility lies with "them," not with "us." At the same time, lack of intellectual and emotional commitment by governments and local people would provide a framework in which the chances of failure are increased.

Are there then, one is moved to ask, tested ways and means by which these governments can produce the necessary changes, and what would be the elements of an appropriate program?

A common and inconvenient characteristic of needed steps is that they are long range. A tractor, a turbine, a generating station, and even a major engineering feature on a river can be purchased, designed, and built in more or less short order and in a predictable time pattern. Financing is a bottleneck that any funding organization can remove. There is both precedent and talent. Changes in values and organization, on the other hand, are not for sale, nor are there patterns to follow. Their emergence is not a matter of funding and their outcome is not predictable in time or in shape. Surely, an educational system can be expanded, its quality improved, its direction changed, and all this can—and should—be financed. But when all is said and done, success, even remote in time, will flow only from altered perceptions, beliefs, and goals. To cite Soedjatmoko again, "in the final analysis self-sustained development and modernization concern the capacity of the entire social system to deal rationally with new problems and challenges." We know little about how this can be accomplished in our own society, and even less outside our range of culture.

The best that those involved in development can do is (1) at all times to be aware of the constraints imposed, (2) to use all possible occasions and avenues to acquire increased knowledge of societal characteristics and their changes—and make the required funds available, and (3) to seek opportunities for setting up, on whatever scale offered, models not only of operating farms and groups of farms, but also of social arrangements that are as novel as is scientific farming and that provide demonstrations of consonance between new achievements in production and human behavior—with care to link the latter to values that have prevailed in the past.

Given these complexities, international organizations no less than donor countries must beware lest their desire to see economic development proceed along certain lines lead them into assuming a primary responsibility which really is not theirs. At times it may even be better to see development falter than to see primary responsibility pass out of the hands of the riparian countries. By the same token any line of action that spells increased understanding and participation of people in the region, of local groups and organizations, and of governmental units should receive high priority in assistance. For only to the degree that the people in the Basin make the plans for development theirs will they succeed in bringing development worthy of the name.